Nanotechnologies for Future Mobile Devices

Learn how nanotechnologies, mobile communicat' each other, and explore the potential for nanotechi and Internet communications and the value network: Based on a research collaboration between Nokia, Heisiiki and the University of Cambridge, here leading researchers and business analysts review the current state-of-the-art and future prospects for:

- Structural materials in mobile devices, including novel multifunctional materials, dirt-repellent, self-healing surface materials, and lightweight structural materials capable of adapting their shape.
- Portable energy storage using supercapacitor-battery hybrids based on new materials including carbon nanohorns and porous electrodes, fuel cell technologies, energy harvesting, and more efficient solar cells.
- Electronics and computing advances reaching beyond IC scaling limits, new computing approaches and architectures, embedded intelligence, and future memory technologies.
- Nanoscale transducers for mechanical, optical, and chemical sensing, nature's way of sensing and actuation, biomimetics in sensor signal processing, and nanoscale actuation.
- Nanoelectronics, for example based on graphene, to create ultrafast and adaptive electronics for future radio technologies, such as cognitive radio.
- Flat panel displays – how nanotechnologies can be used to achieve greater robustness, improved resolution, brightness and contrast, as well as mechanical flexibility.
- Open innovation in nanotechnology development, future manufacturing, and value networks.
- Commercialization of nanotechnologies.

Tapani Ryhänen is Laboratory Director of Nokia Research Center, heading Nokia's research laboratories in Cambridge and Lausanne. He is also responsible for Nokia's research collaboration with the University of Cambridge and the Ecole Polytechnique Fédérale de Lausanne (EPFL), and he is a Visiting Lecturer at the University of Cambridge.

Mikko A. Uusitalo is Principal Member of Research Staff at Nokia Research Center, Helsinki, where he has worked since 2000. He is a Founding Member of the Wireless World Research Forum (WWRF) and EUREKA CELTIC Initiative, and he is a Senior Member of the IEEE.

Olli Ikkala is Academy Professor of the Academy of Finland and Professor of Polymer Physics and Molecular Nanostructures at the Helsinki University of Technology. His current research interest is in developing concepts for functional nanomaterials based on macromolecules, synthetic polymers, and biomacromolecules.

Asta Kärkkäinen is Principal Scientist at Nokia Research Center, Helsinki, where she has worked since 1997. As a Research Manager of the Predictive Engineering Group, she has coordinated research on multidisciplinary design to enable optimal structural, thermal, acoustical, and chemical performance of future Nokia devices.

Nanotechnologies for Future Mobile Devices

Edited by

TAPANI RYHÄNEN
Nokia Research Center, Cambridge

MIKKO A. UUSITALO
Nokia Research Center, Helsinki

OLLI IKKALA
Helsinki University of Technology

ASTA KÄRKKÄINEN
Nokia Research Center, Helsinki

CAMBRIDGE
UNIVERSITY PRESS

CAMBRIDGE
UNIVERSITY PRESS

University Printing House, Cambridge CB2 8BS, United Kingdom

One Liberty Plaza, 20th Floor, New York, NY 10006, USA

477 Williamstown Road, Port Melbourne, VIC 3207, Australia

314-321, 3rd Floor, Plot 3, Splendor Forum, Jasola District Centre, New Delhi - 110025, India

103 Penang Road, #05-06/07, Visioncrest Commercial, Singapore 238467

Cambridge University Press is part of the University of Cambridge.

It furthers the University's mission by disseminating knowledge in the pursuit of education, learning and research at the highest international levels of excellence.

www.cambridge.org
Information on this title: www.cambridge.org/9780521112161

© Cambridge University Press 2010

First published 2010

A catalogue record for this publication is available from the British Librar

ISBN 978-0-521-11216-1 Hardback

Additional resources for this publication at www.cambridge.org/9780521112161

Contents

List of contributors

Gehan Amaratunga
University of Cambridge

Piers Andrew
Nokia Research Center

Marc Bailey
Nokia Research Center

Alan Colli
Nokia Research Center

Tom Crawley
Spinverse Ltd

Vladimir Ermolov
Nokia Research Center

Andrew Flewitt
University of Cambridge

Christian Gamrat
CEA-LIST

Markku Heino
Nokia Research Center

Olli Ikkala
Helsinki University of Technology

Laura Juvonen
Spinverse Ltd

Risto Kaunisto
Nokia Research Center

Jani Kivioja
Nokia Research Center

Pekka Koponen
Spinverse Ltd

Asta Kärkkäinen
Nokia Research Center

Finbarr Livesey
University of Cambridge

William Milne
University of Cambridge

Tim Minshall
University of Cambridge

Letizia Mortara
University of Cambridge

Johann Napp
University of Cambridge

Pirjo Pasanen
Nokia Research Center

Aarno Pärssinen
Nokia Research Center

Markku Rouvala
Nokia Research Center

Tapani Ryhänen
Nokia Research Center

Yongjiang Shi
University of Cambridge

Mikko A. Uusitalo
Nokia Research Center

Di Wei
Nokia Research Center

Yufeng Zhang
University of Cambridge

Preface

Human culture is simultaneously extending its capabilities to master the physical world at its molecular scale and to connect people, businesses, information, and things globally, locally, and pervasively in real time. Nanotechnologies, mobile communication, and the Internet have had a disruptive impact on our economies and everyday lives. Nanotechnologies enable us to use physical, chemical, and biological processes to create new functional materials, nanoscale components, and systems. This book explains how these technologies are related to each other, how nanotechnologies can be used to extend the use of mobile communication and the Internet, and how nanotechnologies may transform future manufacturing and value networks.

At the beginning of 2007, the University of Cambridge, Helsinki University of Technology, and Nokia Research Center established a collaboration in nanotechnology research according to open innovation principles. The target has been to develop concrete, tangible technologies for future mobile devices and also to explore nanotechnologies in order to understand their impact in the bigger picture. The collaboration is based on joint research teams and joint decision making. We believe that this is the proper way to build a solid foundation for future mobile communication technologies. The book is based on the visions of researchers from both academia and industry.

During the summer of 2007 a team of researchers and industrial designers from the University of Cambridge and Nokia created a new mobile device concept called Morph. The Morph concept was launched alongside the "Design and The Elastic Mind" exhibition at the Museum of Modern Art (MOMA) in New York, has been featured in several other exhibitions, won a prestigious *reddot* design concept award, and has had considerable publicity – especially in the Internet. To date, the concept has been viewed over three million times on YouTube. The story of Morph illustrates how nanotechnologies are linked to our everyday artifacts and our everyday lives. In our messages we have always emphasized realism and the responsible introduction of these new technologies to future products. We need to understand thoroughly both the opportunities and risks.

The public interest in the Morph concept may be related to the concreteness of everyday nanotechnology applications illustrating tangible, appealing consumer benefit and value. If the story of Morph was directed to a wider audience, this book is targeted at researchers and people creating future technology and business strategies in both industry and academia. However, we still emphasize the two issues, concreteness and consumer value. Our target has not been to write a comprehensive textbook or a review of nanotechnologies for future mobile devices but through selected examples to illustrate the impact on key mobile device technologies, manufacturing, value networks, innovation models, and ultimately on human societies. Our approach is also critical: sometimes the impact of a new technology is not straightforward and needs to be evaluated against competing technologies that may already be commercially available.

This is a vision statement of academic and industrial researchers working together in the spirit of open innovation. We hope that our book helps to promote stronger links between people working in different fields creating future concepts of mobile communication, Internet services, and nanotechnologies.

1   When everything is connected

T. Ryhänen, M. A. Uusitalo, and A. Kärkkäinen

1.1   Introduction

1.1.1   Mobile communication and the Internet

The Internet has created in only one decade a global information network that has become the platform for communication and delivering information, digital content and knowledge, enabling commercial transactions and advertising, creating virtual communities for cocreating and sharing their content, and for building various value adding digital services for consumers and businesses. The Internet phenomenon has been a complex development that has been influenced by several factors – an emerging culture that shares values that are brilliantly summarized by Manuel Castells [1]:

The culture of the Internet is a culture made up of a technocratic belief in the progress of humans through technology, enacted by communities of hackers thriving on free and open technological creativity, embedded in virtual networks aimed at reinventing society, and materialized by money-driven entrepreneurs into the workings of the new economy.

The Internet can be characterized by four key elements: Internet technology and its standardization, open innovation based on various open source development tools and software, content and technology creation in various virtual communities around the Internet, and finally on business opportunities created by the Internet connectivity and access to the global information. The history and the origin of mobile communication are different and have been driven by the telecommunication operators and manufacturers. Digital mobile communication has focused on providing secure connectivity and guaranteed quality of voice and messaging services. The key driver has been connection, i.e., establishing a link between two persons. The global expansion of digital mobile phones and mobile network services has occurred in a short period of time, more or less in parallel with the Internet, during the last 10–15 years. Today there are roughly 3 billion mobile subscribers, and by 2010 nearly 90% of the global population will be able to access mobile voice and messaging services.

Mobile communication networks have also evolved from the original voice and text messaging services to complex data communication networks. The mobile phone

Nanotechnologies for Future Mobile Devices, eds. T. Ryhänen, M. A. Uusitalo, O. Ikkala, and A. Kärkkäinen. Published by Cambridge University Press. © Cambridge University Press 2010.

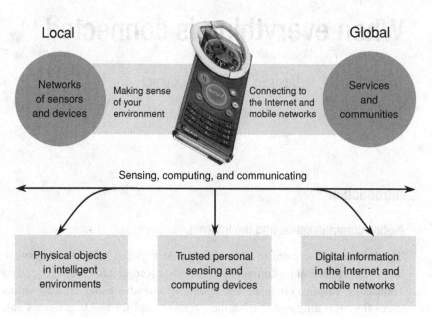

Figure 1.1 Mobile device as a gateway between local and global information and services.

has become a pocket-size mobile multimedia computer with various applications and capabilities to access networked services. The convergence of mobile communication and the Internet is one of the most signicant technology trends of our time with signicant social and economic impact. The mobile devices and networks are able to extend access to the Internet from homes and offices to every pocket and every situation in everyday life. Mobile phones outnumber personal computers by a factor of 5–10. In developing countries in particular, mobile phones are immensely important for accessing the Internet and its services as many people make their first connection to the Internet via a mobile phone.

It is very clear that the evolution of human society will be shaped by the global communication and information sharing networks. In this book we will discuss another important dimension that will bring the Internet and mobile communication even closer to human everyday life. We will discuss the new capabilities to master our interface to the physical world. The ability of human technologies to image, measure, and manipulate matter down to the molecular scale and to master the self-organizing structures of nature will extend our ability to invent new materials and manufacturing solutions, new energy, sensing, computing and communication technologies, and a deeper means to interact with living systems: our environment and our bodies. The aim of this book is to study these abilities in the context of the Internet and mobile communication and from the perspective of the era of the transformation of human society towards the new concept of seamless local and global interaction, illustrated in Figure 1.1.

1.1.2 Towards merging of physical and digital worlds

Mobile phones have already become an enabling platform for several digital services and applications. Mobile phones are now mobile computers with a wide range of multimedia functionality, e.g., imaging, navigation, music, content management, internet browsing, email, and time management. Increasingly they will have advanced multiaccess communication, information processing, multimedia, mass storage, and multimodal user interface capabilities.

Mobile phones are developing towards being trusted personal intelligent devices that have new fundamental capabilities, illustrated in Figure 1.1:

- to sense and interact with the local environment via embedded short-range radios, sensors, cameras, and audio functionality;
- to function both as servers for global and local internet services and as clients for global internet services;
- to serve as gateways that connect local information and global Internet-based services;
- to carry the digital identity of the user and to enable easy-to-use secure communication and controlled privacy in future smart spaces;
- to make sense of and learn from both the local context and the behavior of its user, and optimize its radio access, information transport, and device functionality accordingly.

Form factors and user interface concepts of mobile phones and computers will vary according to the usage scenario. The trend towards smaller and thinner structures as well as towards reliable transformable mechanics will continue. The desire to have curved, flexible, compliant, stretchable structures and more freedom for industrial design sets demanding requirements for displays, keyboards, antennas, batteries, electromagnetic shielding, and electronics integration technologies. Integrating electronics and user interface functions into structural components, such as covers then becomes a necessity.

The modular device architecture of mobile phones and computers consists of several functional subsystems that are connected together via very high-speed asynchronous serial interfaces [2, 3]. This modular approach enables the use of optimal technologies for particular functionalities, optimization of power consumption, and the modular development of device technologies and software. The same modular architecture can be extended from one device to a distributed system of devices that share the same key content, e.g., remote mass storage, display, or a printer.

A variety of new devices will be embedded in our intelligent surroundings. Ambient intelligence will gradually emerge from the enhanced standardized interoperability between different consumer electronics products and will extend into more distributed sensing, computing, storage, and communication solutions. The current communication-centric modularity will develop into content- and context-centric device virtualization.

The vision of ambient intelligence, in which computation and communication are always available and ready to serve the user in an intelligent way, requires mobile devices plus intelligence embedded in human environments: home, office, and public places. This results in a new platform that enables ubiquitous sensing, computing, and communication. The core requirements for this kind of ubiquitous ambient intelligence are that devices are autonomous and robust, that they can be deployed easily, and that they survive without explicit management or care. Mobility also implies limited size and restrictions on the power consumption. Seamless connectivity with other devices and fixed networks is a crucial enabler for ambient intelligence systems. This leads to requirements for increased data rates of the wireless links.

Intelligence, sensing, context awareness, and increased data rates require more memory and computing power, which together with the limitations of size lead to severe challenges in thermal management. It is not possible to accomplish the combination of all these requirements using current technologies. As we shall see in the rest of the book, nanotechnology could provide solutions for sensing, actuation, radio, embedding intelligence into the environment, power-efficient computing, memory, energy sources, human–machine interaction, materials, mechanics, manufacturing, and environmental issues.

1.2   Future devices, lifestyle, and design

1.2.1   Navigation in space and time

Early on the morning of Tuesday July 7, 2020, Professor Xi wakes up in her modern Kensington hotel room in London, still feeling tired after her long flight from Shanghai. Only a few years earlier she limited her traveling to the most important conferences and visiting lectureships. Her environmental ethics and the existence other means of communicating meant she preferred to stay mostly in her own university in Hangzhou. However, this invitation to give a series of lectures about Chinese innovation and environmental strategies to the MBA class of the London Business School gave her an opportunity to discuss both the history of innovation in China and its recent huge economical and technological advances.

After having breakfast Professor Xi decides to use the remaining hour before her meeting with Professor Williams to take a walk through a London park. It is a beautiful morning. Walking through the city, Professor Xi relies on her mobile device. Her thoughts return to her lectures. She opens her Morph device which shows her a map of the city, her location, and the route. She lifts the device, looks through it, and the device displays the street and the local information on services around her (see Figure 1.2). This is the modern equivalent of navigation, cartography, measuring of distances, and using a compass. All these ancient inventions of the Song dynasty are now encapsulated in this transformable piece of flexible and elastic material. She wraps the device around her wrist and walks on to her meeting, trusting in the instructions of her mobile personal device.

Professor Xi is moving through the streets of London that are depicted as a mixed experience of physical and virtual realities. At the same time she is able to find information about her environment that helps her to navigate towards her meeting. She is able to access local information about her surroundings – directly based on the local

Figure 1.2 A transparent and transformable user interface [4] for augmented reality experience.

short-range communication and indirectly using the global Internet-based knowledge relating to her location. Location-based information services have thus become an integral part of living in a physical space. This access to local knowledge enables her to interact with the people, commerce, history, and the aspirations of the immediate community of what otherwise would remain merely a nondescript street to her. It helps her to link her current physical location to her personal world map.

Her personal mobile wearable device is a physical expression of the *timeless time* and the *spaces of flow* [1, 5, 6]. The device connects her to her physical location but at the same time also to the global information that describes and defines her location. The space and time of her experience have become much more complex and nonlinear. However, she also has a physical meeting to attend, and the device gives her instructions for how to get there in time.

1.2.2   Transformable device

A transformation is defined in mathematics as a process by which a figure, an expression, or a function is converted into another that is equivalent in some important respect but is differently expressed or represented. In a similar way a transformable device has its own character but its functional and morphological appearance can be adjusted according to the context. In Figure 1.3 Professor Xi's mobile device, the Morph, presents a potential future map with navigation functionality. The same device can be opened to form a flat touch display that shows a dynamic and context aware map of the environment. The dynamic map combines space and time, it can relate current events and the history of its local surroundings, guiding us through space and time. The same device can be folded into a handheld communication tool that enables us to easily find the people and things that we need in our current situation. Furthermore, the device can be wrapped around the wrist so that it becomes a simple street navigator guiding us to our goal.

Figure 1.3 A future guide could be a device that transforms from a wearable navigator to a touch-sensitive map.

Morph has some new capabilities that are not possible with existing technologies: it is a flexible and stretchable device made of transparent materials with embedded optical and electronic functions. To achieve this we need:

- a transparent device with display capability;
- flexible and partly stretchable mechanics with nonlinear spatial and directional control of elasticity embedded into the materials themselves and even rigid-on-demand actuators;
- distributed sensors and signal processing in the transparent structures, e.g., pressure and touch sensor arrays;
- transparent and flexible antenna, electronics, and energy storage;
- externally controllable and dynamic surface topography and roughness;
- multifunctional, robust surface coatings providing protection of device functionality, dirt repellence, antireflection, etc.;
- transformability and conformability with intelligence that can extract conformation and context and adjust the functionality accordingly.

The Morph device can essentially be transformed in many different ways: e.g., into a graphical user interface, or a mechanical configuration to increase the availability of applications and services. The user interface of the device can adapt to the needs of the user in terms of both its functionality and its appearance. Transformability can be used to enable ease of use of the device, applications, and services. The Morph device is transformable in both its form and its conformation. The Morph is a cognitive user interface, capable of sensing both the user and the environment, making decisions based on this information, adapting to the context and giving feedback to the user. The Morph can learn about its user and can become a trusted personal companion.

The transformable compliant mechanisms need to be built deep into the material solutions of the device. In it, complex mechanical and electromagnetic metamaterials and artificial nanoscale material structures enable controllable flexure and stretch in the macroscopic mechanisms creating the desired functions. One of the key challenges is how to freeze the conformations and how to allow the right level of flexibility in various parts of the device. Can nanostructured functional materials be arranged into macroscopic structures according to the device functionality? We need to learn how to build a bridge between the nanoscale properties of the materials and the manufacturing of a macroscopic object.

1.2.3 Fashion and expression

The mobile device has already become a personal device that is used to express personal identity. New materials, integration technologies, modular architectures and tools of customization will affect the way people can design and tailor their devices – according to the desired functionality and the look and feel. The personalization of mobile devices and the integration of some of their functions, such as display and sensors, in clothing will create new opportunities for fashion designers.

Nanotechnologies will enable the integration of electrical and optical functions into fabrics. We are not speaking only about integration of components into wearable objects but also of embedding functions into the materials that are used in clothes and other wearable objects. Manufacturing solutions, such as ink jet printing and reel-to-reel fabrication, have already been used in the textile industry. As electronics manufacturing is beginning to use similar methods, the integration of electronics and optics will become feasible on large scale in textiles and other materials will be used for wearable objects.

New engineered materials and structures could be used to create artificial, functional, and biocompatible systems. For example, early prototypes of artificial skin with integrated sensors have been developed [7, 12]. Artificial sensors can be integrated into the human nervous system. This kind of functional system will be complex and expensive but it is possible to create low-cost functional systems with sensing and actuation: materials that react to temperature, humidity, and touch by changing their properties, such as surface roughness, perforation, color, and reflectivity. Many phenomena that exist in nature could be mimicked by future clothing and wearable devices.

Fashion is used as a way to express both being part of a community and individuality. In the future fashions will also enable us to express our feelings and adapt dynamically to our surroundings. At the same time, our personal, wearable objects will become intelligent and networked to the surrounding virtual and physical worlds.

1.2.4 Lifestyle and the mobile device (global knowledge, local view)

Mobile phones and mobile communication have also had significant impact on human culture. Mobile communication has created the capability for simultaneous social interaction at a distance in *real time*. Mobile connectivity has increased the efficiency of different professionals on the move. Social networks, including the family, are reshaping based on pervasive connectivity. Several public services, e.g., remote health care, can be provided using mobile-communication-based technologies. Mobile phones have created youth cultures: peer groups with a collective identity, strengthened personal identities expressing creativity and fashion, and new patterns for consumption of entertainment, games, and media. Mobile communication can be used to improve the safety and security of people; however, the risks and fears of misuse of the technologies are real: the security and privacy of individuals are also threatened by omnipresent communication technologies.

Internet connectivity and mobile communication have had a great influence on human social and cultural development. At the deepest levels our concepts of space and time are changing. Mobile networks and pervasive connectivity are enablers for commercial transactions and advanced services, even production and manufacturing will become geographically distributed globally. Mobile communication is a most advanced technology that will enable this kind of global network society [6]. Human society is becoming more dependent on knowledge generation and global flows of information than on the physical locations in terms of not only economies, but also cultural trends. Information networks and mobile connectivity are also beginning to affect the physical environments of human beings, our urban planning, and the architecture of our buildings and cities [8]. The capability to connect local and global information in the same physical location are changing our concept of distance and the location itself.

The concept of time has varied throughout history. Contemporary societies are driven by the concept of clock time that is a result of the industrial age. During this era of Taylorian concepts time was modeled as linear, irreversible, measurable, and predictable. The profound transformation of the concept of time in the globally networked society is based on instant information accessibility and interactions that lead to emerging, random, nonlinear, and incursive phenomena that shape the life and economics of the people. Manuel Castells has named this emerging era the timeless time [5]. The timeless time concept can be seen in several current phenomena. The ability of global markets to work *in real time* with transactions occurring at immense speed creates the need for more flexible management processes in networked enterprises in order to make their decisions *in real time* based on understanding complex emerging situations. The increasing speed of the economic activities influences the patterns of work and free time. Even social interaction and social networks obey more complex causalities – the emergence and exponential growth of online communities, such as Facebook, Twitter, and LinkedIn, are reflections of the future of social networks.

Our concepts of macroscopic space and time are changing, and at the same the human capability to study, image, and manipulate matter at a molecular scale is expanding our scale of impact. Nanoscience and nanotechnologies are creating possibilities for developing new functional materials, new sensing and computing solutions, and technologies for environmental and medical engineering. Nanotechnologies are enabling integration of new functionality into mobile communication and the Internet. The Internet will become connected to our physical environment.

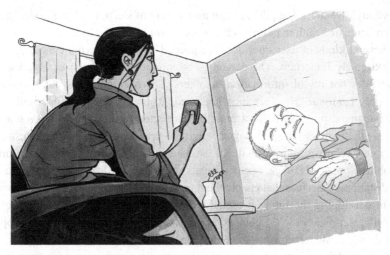

Figure 1.4 Professor Xi waits for the ambulance with her father. She is in a visitor room at the London Business School and he is at home in Shanghai. Their communication is provided by their Morphs and the nearby screens. The prewarning via the Morphs leads to quick preventative action by medical personnel and keeps relatives informed.

1.3   Trusted personal device becomes a cognitive interface

1.3.1   Assisted living and remote health care for the elderly

Professor Xi is drawn from her mental preparations for her lectures by an insistent alarm signal from her Morph. It takes her a moment to realize the cause: the Morph is notifying her that her father in Shanghai is showing signs of potential heart failure! Quickly she uses her Morph and the largest wall screen in the visitor room of the London Business School to make a virtual connection to her father (see Figure 1.4). She and her father feel as if they are in the same room, even though he is in Shanghai. He is not in any pain and so they talk for the 5 minutes it takes for the ambulance to arrive. During the ambulance trip, and in the hospital, Professor Xi continues to reassure her father via their Morphs. Soon after his arrival at the hospital it is clear that her father is not in immediate danger but the hospital wishes to keep him in for several hours for observation. A relieved Professor Xi closes the connection and tries to continue her preparations. She feels grateful that her father's health monitoring system, combined with the quick reactions of the health system, has saved her father from serious harm.

Technologies for assisted living and remote health care for the elderly are urgently needed. With such systems, the increasing elderly population can receive care while staying in their own home and so maintaining social independence. This preventative system could allow significant savings in public expenditure on health care. As the population profiles are changing, we face a truly global challenge in China, Japan, North America, and Europe. The same technologies need not be used solely for the remote monitoring of the elderly, e.g., they could monitor pilots during flights or monitor anybody's wellness, e.g., the amount and quality of sleep, level of physical activity and exercising, or health markers such as cholesterol levels.

Several kinds of user interfaces are needed in assisted living concepts, and privacy is extremely important. The person to be monitored needs a wearable user interface that does not reveal information to others. The user interfaces and devices need to be truly *personal*. The user interface needs to be either compatible with clothing or an integrated part of it, and it must be connected to other devices, such as a mobile device that provides the global connectivity. People with the right to use the information, like near relatives and professional caregivers, could access the information via their mobile devices. Healthcare professionals would benefit from a combination of mobile and stationary devices that enable them to continuously monitor their patients.

In developing the future personal trusted mobile devices the meaning of trust has specific significance in the context of medical applications. A failure in the operation of the device could result in tragic consequences. The secure operation of devices over the Internet and through mobile communication requires intelligent algorithms embedded in devices, networks, and services. The end-to-end solution (according to Figure 1.1) must be able to monitor its own integrity and give warning of possible failures. The devices and the solutions need to pass the specific tests and regulations required for medical equipment. For fitness, wellness, and preventative healthcare purposes the requirements are somewhat less demanding. In addition to diagnosis, active treatments inside the human body are being developed based on bionanoscience. A good example of this is the controlled release of drugs for targeted or long-term delivery.

1.3.2   Integrated cognitive systems

Energy-efficient implementation of artificial sensors on mobile devices and in intelligent environments will enable the recognition of people, objects, chemicals, radio traffic, sounds, speech, touch, and the overall context around the devices. These embedded capabilities and increased artificial intelligence could bring communication, applications, and Internet-based services to a new level with only imagination as the limiting factor.

Integrated cognitive systems and processes are capable of perception, cognition, learning, and actuation in real time in real physical environments. Ultimately these artificial cognitive systems will seamlessly integrate with human cognitive processes and behavioral patterns. Thus the first requirement for embedded intelligence is efficient interaction with humans – the users.

Solutions for artificial classification of sensory information, mechanisms for decision making driving the device's actions, and processes for learning and adaptation need to be implemented in devices with different levels of complexity. If the device can understand the context, behavioral patterns, gestures, and moods of the user, then future devices and user interfaces can change and adapt their operations to support their user accordingly. Going beyond this, embedded intelligence could support us in presenting options, remembering, and even decision making.

Based on efficient human interaction and context awareness the cognitive user interface of our devices could create an enhanced new linkage for everyone to the merged physical and digital world in which we live. Our personal, trusted mobile device will then take on an essential role – connecting us and acting as the digital agent that participates in the actions of the local physical and digital environment. In our vision, the mobile device would become a seamless part of the surrounding ecosystem of resources for sensing, computing, local memory, and global communication.

1.4 Ambient intelligence

1.4.1 Augmented reality

Professor Xi is advising her students in China from her hotel room in London. Nowadays she always meets her students in the virtual world as they all have their avatars and secure access to the course space. The class room, the laboratories, the analyzed data, and the simulations are always available, while the course space supports ongoing repetition and feedback, making the virtual world a perfect place for communication with the students and a powerful tool for teaching.

After finishing the last preparations for her lectures, Professor Xi has some spare time. She takes her Morph and puts on her virtual glasses with three-dimensional audiovisual capabilities that have been individually tailored for her to give a complete experience of being in a specific space. With a cognitive user interface based on speech recognition, she selects to go virtually to her favorite forest in China. As a child she used to play there with her parents and friends. The forest no longer exists but she and her family can still visit it virtually, alone or together. She closes her eyes and relaxes by just listening to the sounds of the forest around her.

Suddenly Professor Xi remembers that she had planned to go shopping at Harrods. Unfortunately she no longer has time to go there physically, so she goes using virtual reality (see Figure 1.5). While walking through the virtual Harrods, the virtual reality system compares the model of her body to the clothes in stock and determines which will fit her. She can see images of herself wearing these clothes at the places in the store where the clothes are displayed. She quickly finds what she is looking for, pays for the dress, and asks for it to be delivered to her hotel room. By having her own virtual model with her measurements updated whenever needed, her clothes are now a much better fit. After this virtual shopping she closes the connection and hurries to give her lecture – this time in a real physical space.

Products will become more reliable and more suited to individual needs. The end user will be connected to producers via the Internet. The producer would then follow the consumer trends and needs in real time and even tailor their products to specific customers. Products can be tracked and monitored during manufacturing and transport – or throughout their entire lifecycle. Even the quality of the products and services will improve. The possibility of tracking products will give consumers the option to identify the ethical origin of the product. It will be possible to check the origin, health factors of food, and the green footprint of different food suppliers. Virtual communities of consumers will be able to increase the consciousness of possible misuses and advise

Figure 1.5 Buying clothes is easier with the help of a computer aided design (CAD) model of the body. It can be used while visiting virtual stores, but it can also be helpful while shopping in a real store, e.g., it can be used to study how clothes might fit without putting the clothes on!

each other how to live wiser, safer, and healthier lives but regulation will be needed to guarantee the validity and reliability of the product information.

A mobile personal device is a natural gateway between the physical and digital worlds and a user interface that combines them into one experience – an augmented reality. These mixed reality places can be either open to all or protected behind a secure and limited access. Combining physical and digital domains carries substantial risks for physical safety, security, and privacy. New solutions based on technologies, such as quantum cryptography, chemical tags, DNA-based identification, are needed to solve these challenges.

1.4.2 Embedded ambient intelligence

The vision of pervasive, ubiquitous computing or ambient intelligence has been the subject of large-scale research in Europe, USA, and Japan. There exist different technologies and concepts for its implementation but all these refer to intelligent, artificial environments based on sensors, processors, memory, and radios that are sensitive and responsive to people. We have already discussed most of the key attributes: embedded devices, personalized services, context aware user interfaces and services, adaptive and anticipatory processes. We have discussed these elements from the perspective of the mobile device. However, the implementation of ambient intelligence requires a much larger number of smaller devices that are embedded in our physical environment.

The connectivity between physical objects and the Internet has been studied under the concept of the *Internet of Things* [16]. The previously discussed applications and services of assisted living, augmented reality, and e-commerce require new ways of connecting physical objects to the Internet and to local sensor networks. Today we are already capable of adding autonomous intelligence to handheld devices; however, the requirements for sensors and other embedded devices in ambient intelligence are more challenging: they must be robust, autonomous in terms of operation and energy, self-configuring, and easy to deploy. Here the emerging capabilities of nanotechnologies are creating new options.

In the future everything will be connected. This means that we will have tools for organizing our life in a new, intelligent way. Embedded intelligence will be applied in industrial systems, private spaces, and public infrastructure, e.g., for optimizing transport and saving energy. We will get new tools to structure our life, novel methods, and more design space to optimize complex systems. We can ask if it is possible to design better environments for people by distributing intelligence and centralizing all kinds of energy-hungry technology, e.g., everything that requires movement against gravitation. Billions of mobile devices, including sensors, connected to each other and to servers, will result in an enormous wireless sensor network. By collecting the sensor data from volunteers (and by giving something back, e.g., the resulting data) we can map, e.g., the distribution of pollutants or the transmission of diseases.

Using materials that are functional at different scales, we can build systems to tune and customize the properties of the access points to the smart spaces. They would not only enable the devices to adapt to the environment in a secure way, but they could also be used to enhance human capabilities. We can already read radio frequency identification (RFID) tags, but in future we could have devices that will enable us to see in the dark, hear high-frequency sound, hear in noisy places, and distinguish things by their chemical signatures using artificial olefactory capabilities beyond human reach.

The main challenges for building the embedded electronics systems of the future lie in the design of complex electronic systems. How to integrate features? How to treat the interfaces between different scales? How to do real time sensing and computing? How to make them safe? Which parts will be done using software and which parts will be done using hardware in order to make our embedded system interoperable, accessible, distributed, sensorial, wireless, energy- and cost-efficient?

1.5   Technology challenges for humankind

1.5.1   Looking for the global solution

"Someone has said: 'We do not own this country – we borrow it from our children'. I hope that I nearly always remember this principle when I am making my personal everyday decisions. We all need clean air, pure water, food, and shelter against cold

Figure 1.6 "The networked world is the key enabler to treat global human challenges and to leverage the opportunities of novel technologies such as nanotechnologies, information technologies, cognitive science, and biotechnology," Professor Xi lectures, illustrating her presentation via her Morph.

weather and rain. This requires energy, and we consume natural resources. I wonder if I really know what is best but I have started trusting in the information that is available from my virtual community which helps me to follow green footprints that are based on valid data of different products and services," Professor Xi opens her lecture in her modest way.

Standing in front of her students in the modern classroom in London (see Figure 1.6) she begins to lead her students towards the new sustainable, distributed, and networked concepts of production of goods and services. She continues: "Products and services belong together. We are globally creating solutions that provide healthcare, education, quality of life but also key physical elements like pure water and food, clean air. We are today producing our goods closer to the end-consumers, tailoring the products to their true needs. The networked world is the key enabler to address global human challenges and to leverage the opportunities of novel technologies such as nanotechnologies, information technologies, cognitive science, and biotechnology."

1.5.2   Global population is ageing and polarizing

The global rate of growth of the human population is faster than ever. On the other hand, the population in the industrialized world has stabilized, and the average age is rapidly increasing. In many of the developing areas of Asia, Latin America, and especially Africa, population growth is leading to an unsustainable situation with several consequences. The rapidly growing urban population of the less developed world will create a challenge that does not have any clear and easy solution.

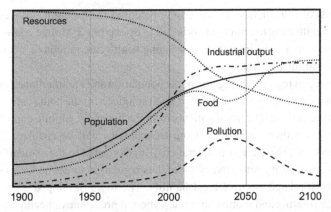

Resources

Industrial output

Food

Population

Pollution

1900   1950   2000   2050   2100

Figure 1.7 Simulation of the world state according to the *Limits-to-Growth Model* [10]. The model is based on the assumption of global success in stabilizing the population and the industrial output per person and creating efficient technologies for pollution control, sustainable production, and agriculture.

The trend in the human population towards the year 2050 shows several important features according to United Nations statistics. The aging population and the lack of skilled labor are consequences of the saturation of population growth in the more developed regions. However, the rapidly growing urban population in the less developed regions is the most severe challenge for the future of the world, leading directly to social polarization along the lines of poverty, misery, migration on various spatial scales, criminality, corruption, famine, and wars. The rapidly growing elderly population is also a challenge in some less developed regions, especially China. Some population polarizations have become extreme – the global population is simultaneously suffering from obesity and famine. The challenges in the rural regions of the world are related to the overharvesting of tropical soils and forests and the lack of clean freshwater. Population growth is fundamentally linked to the fast decline of natural resources, food, and water quality, and to increasing pollution, as shown in even the most optimistic scenario of the Limits-to-Growth Model [10] depicted in Figure 1.7.

The Internet and mobile communication will have an essential role in finding solutions to global population challenges. A large part of the urban population in the less developed regions will be constantly moving and living under very unstable conditions with limited infrastructure. The mobile phone and its services will be the enabler for bringing information, knowledge, and education to these people. The mobile phone will also help these people to keep in contact with their families and friends. Communities of people can be created with easy-to-use mobile Internet services. Education and knowledge are the key to improving the quality of life and to solving different environmental problems. Furthermore, individualization of labor in less developed regions will create a demand for solutions for commercial transactions and managing small businesses.

The rural areas of less developed regions will access the Internet and its services via mobile communication and mobile phones. Already today mobile phones and text messaging are becoming part of everyday life in rural areas in terms of communicating and accessing vital information. However, the potential for bringing services, such as education and health care, to remote areas is still to be exploited. Mobile communication technologies are going to be important in bringing health care to remote areas with less developed medical infrastructure.

In more developed regions, the ageing of the population and the simultaneous decrease in labor in medical care increase the need for good solutions for the home care of disabled people and for the early discharge of patients from the hospitals. Mobile communication and the Internet together with smart home technologies will be enablers for remote health care. However, there are still several challenges related to the standardization of data formats, data security, and privacy, as well as regulation to define levels of system reliability and the related issue of legal liability. The more developed areas will also benefit from the Internet and mobile communication in preventative health care, such as fighting against obesity.

1.5.3   Environmental challenges need global solutions

Environmental problems like global warming require focused investments and new solutions. In order to have more than five billion people living in peace together while saving energy, we need the best applications that technology can offer. We have to find solutions to questions like: How do we live in future? What kind of cities can we have? How do we travel? How do we manufacture our products? How do we get food and clean water for everyone? How do we get the energy required? How do we save energy? How do we save the planet?

Figure 1.7 refers to simulations based on the famous *Limits-to-Growth* Model created in 1974 by the MIT research group which have been updated using existing data [10]. Figure 1.7 shows the most optimistic scenario that assumes progress in controlling population growth, pollution, and the depletion of natural resources together with new sustainable solutions for industrial and agricultural production. The time window is narrow: solutions are needed within the next 10–20 years. When we think about the combination of the Internet, mobile communication, and nanotechnologies, we have the tools in our hands to create efficient and sustainable solutions.

Today global energy consumption is growing at a very high rate, up to 50% by 2030 [19]. Without major technological changes, fossil fuels will still provide 81% of primary energy demand by 2030 [20]. Energy is important for the basic needs of people: clean water, food, and shelter from cold and rain. It is important for our well being and for enhancing the quality of life. A functioning society and economical development require affordable energy. High energy costs can lead to inflation and environmental problems, like pollution, acid rain, and greenhouse gases. The effect is global: every country is affected, thus international solutions and policy processes, such as the Kyoto Protocol, the Asia–Pacific partnership and the G8 summits, are important. Social stability depends on the availability of resources.

Alternative and renewable energy sources are becoming economically feasible, and the motivation to develop them is clear. However, in parallel we need to enhance the efficiency of our processes, reducing energy consumption and the use of other scarce natural resources. The development of affordable alternative energy sources will be a driver for nanotechnologies, and the efficiency of our processes can be improved by the intelligence that we can build into our environments and artifacts.

Carbon dioxide emissions are far from being under control. The global temperature is warming fast. The Arctic ice cap has declined in size by about 10% since the late 1990s. Climate change and energy are coupled to other future challenges such as having sufficient water resources. As much as 75% of the world population could face freshwater scarcity by 2050 [19, 21]. The water–food–energy supply per person will decrease. International collaboration is needed to meet all these challenges, e.g., climate change, and supplying basic human needs and security.

Ambient intelligent and mobile devices can be used to collect and share information about the global environmental state. Even volunteers with energy-efficient devices equipped with sensors can participate in collecting global and local information.

1.5.4 Sustainable economies and efficient ways of living

We can ask if it is possible to design a better environment for people by distributing intelligence and centralizing all energy-intensive processes. The use of enhanced communication tools, like our mobile gateway device, would reduce the need for transportation and traveling. Our various artifacts can be designed and manufactured to be sustainable. Environmentally friendly materials can be used. The structures and components of the devices can be designed to be easily recyclable. Post-processing of materials and products can be brought nearer to people, in terms of local manufacturing skills, such as three-dimensional printers that enable us to design artifacts and to reuse the same raw material several times.

Novel manufacturing and fabrication methods related to nanotechnology may be key enablers for future sustainable electronics. Printed electronics and related reel-to-reel manufacturing may be the first disruptive solutions (defined according to Christensen's theory of disruptive innovation, see [13]) that enable a new kind of electronics industry: environment friendly, low-cost, large-area electronics that will open new applications in embedding electronics into human environments. Even scaling up from laboratories to the industrial scale is challenging, the bottom-up self-assembling processes can further simplify electronics manufacturing and lower the investments and assets needed to establish manufacturing sites.

We can develop new electronics materials that are easier to recycle and maybe even biodegradable. We can optimize and minimize energy consumption in manufacturing future materials and products. We can create solutions based on nanomaterials that help us to solve the environmental challenges that we will face during the coming decades.

Let us set the right targets, focus on the right set of technologies, and introduce nanotechnologies into the public arena in a responsible way. Nanotechnologies in combination with the global information society are key parts to building a sustainable future.

1.6   About this book

This book discusses possible nanotechnological solutions to the challenges outlined in this chapter. Each subsequent chapter illustrates relevant paths towards future devices by building and extrapolating from a review of the current state-of-the-art.

Chapter 2 reviews the potential applications of structural materials within mobile devices, such as novel multifunctional materials for structural applications and user interfaces, lightweight structural materials, tough and strong materials capable of adapting their shape, and dirt repellent, self-cleaning, and self-healing materials. The aim is not to give a comprehensive list of all relevant materials, topics, and references related to developing Morph-type devices, but rather to discuss the approach whereby nanotechnology-related efforts could answer these needs and then give some topics and current results as examples. The underlying vision is that nanotechnology is expected to provide an extensive toolkit enabling tailoring of materials at the molecular level to generate precise structures with totally new and well-defined functions. This bottom-up approach utilizing self-assembly of selected molecules and nanoscale building blocks forms the basis for the development of novel specific materials.

Chapter 3 investigates how nanotechnology can answer the increasing need for portable energy storage in terms of energy density, environmental friendliness, low cost, flexible form, and fast charging times. Some examples of how these challenges are targeted are supercapacitors, new materials such as carbon nanohorns and porous electrodes, fuel cell technologies, energy harvesting, and more efficient solar cells.

New possibilities for computing and information storage are at the heart of future generations of electronics. New challenges are targeted to be answered by energy efficiency, speed of operation, the capability to embed intelligence in everyday objects, low cost, and the capability to solve new kinds of problems due to efficient implementation of nontraditional computational logic. Chapter 4 reviews the ultimate limits of scaling, new approaches for computing in terms of computational principles, data representation, physical implementation, new architectures, and future memory technologies. Future opportunities are compared in terms of their appropriateness for future mobile devices.

Ubiquitous sensing, actuation, and interaction are needed to allow efficient implementation of the mixed reality of a ubiquitous Internet in contact with the physical world. Chapter 5 reviews sensors for the future, nature's way of sensing, sensor signal processing, the principles of actuation in nature and artificial systems, nanoscale actuation, and the path towards future cognitive solutions.

A central component of past, present, and future mobile devices is radio, enabling wireless communications. New forms of mobile communication, e.g., location-based services, will increase the need for enhanced performance in connectivity. In future, everything – both people and things – will be connected through embedded intelligence. The first step towards the Internet of Things is to enhance the usability of current devices by better connection to the services. The new communication paradigm of cognitive radio will embed intelligence not only in the application level but also in the radio connectivity. Near-field communication based on RFID tags, which in the future will probably be produced using printed electronics, will be the first step towards smart spaces. Cost, size, and extremely low power consumption are the main drivers for low-end connectivity for personal area networks, RFIDs, and other short-range communications. Chapter 6 reviews the principles of radio communication systems, wireless communication concepts, radio system requirements, implementation requirements, and nanoelectronics for future radios.

Chapter 7 examines the history of display technology with reference to mobile devices, considers current technology developments that aim to move displays towards greater robustness, improved resolution, higher brightness, enhanced contrast and mechanical flexibility, and looks forward to the problems that future display manufacturers will need to consider.

Chapter 8 discusses the changing ways by which value can be created and captured from nanotechnologies for future mobile devices. These issues are discussed from three perspectives. Firstly, attention is focused on how advanced technologies emerging from public and private laboratories can be transformed into products and services. Secondly, a look is taken at how new, open models of innovation are emerging and the opportunities and challenges that this new approach presents for firms are discussed. Thirdly, the changing ways in which manufacturing activities are coordinated are examined, in particular, the emergence of value creation networks as illustrated by changes in the mobile telecommunications sector. The chapter concludes by drawing together the key themes discussed and focusing on the implications for value creation opportunities for nanotechnologies applied to future mobile devices.

Chapter 9 discusses the commercial status of nanotechnologies using the Gartner hype curve, analogies with Internet development, and existing measures, such as Internet visibility, patenting, publishing, and stock market trends.

The book ends with our conclusions.

References

[1] M. Castells, *The Internet Galaxy, Reflections on the Internet, Business, and Society*, Oxford University Press, 2001.

[2] www.nota.org

[3] www.mipi.org

[4] Nokia Morph is a mobile device concept created by Nokia Research Center and the University of Cambridge, Nanoscience Centre. Members of Nokia design team were Jarkko Saunamäki, Tapani Ryhänen, Asta Kärkkäinen, Markku Rouvala, Tomi Lonka, Teemu Linnermo, and Alexandre Budde. Members of the University of Cambridge design team were Stephanie Lacour and Mark Welland. First published in P. Antonelli, *Design and the Elastic Mind*, The Museum of Modern Art, New York, 2008. Web page: http://www.nokia.com/A4852062

[5] M. Castells, *The Rise of the Network Society*, second edition, Blackwell Publishing, 2000.

[6] M. Castells, M. Fernández-Ardèvol, J. Linchuan Qiu, and A. Sey, *Mobile Communication and Society, a Global Perspective*, MIT Press, 2007.

[7] S. P. Lacour, J. Jones, S. Wagner, T. Li, and Z. Suo, Stretchable interconnects for elastic electronic surfaces, *Proc. IEEE*, **93**, 1459–1467, 2005.

[8] W. J. Mitchell, *ME++, The Cyborg Self and the Networked City*, MIT Press, 2003.

[9] D. Kennedy and the editors of *Science, Science Magazine's State of the Planet 2006–2007*, AAAS, 2006.

[10] D. Meadows, J. Randers, and D. Meadows, *Limits to Growth, The 30-Year Update*, Chelsea Green Publishing Company, 2004.

[11] W. Weber, Ambient intelligence – industrial research on a visionary concept , in *Proceedings of the 2003 International Symposium on Low Power Electronics and Design, ISLPED '03*, ACM, 2003, pp. 247–251.

[12] G.-Z. Yang, ed., *Body Sensor Network*, Springer, 2006.

[13] C. M. Christensen, S. D. Anthony, and E. A. Roth, *Seeing What's Next*, Harward Business School Press, 2004.

[14] S. E. Frew, S. M. Sammut, A. F. Shore, *et al.*, Chinese health biotech and the billion-patient market, *Nature Biotech.*, **26**, 37–53, 2008.

[15] S. Haykin, Cognitive radio: brain-empowered wireless communications, *IEEE on SAC*, **23**, 201–220, 2005.

[16] ITU report, "The Internet of Things", The World Summit on the Information Society, Tunis, 160–18 November, 2005.

[17] J. Mitola III, *Cognitive Radio Architecture–The Engineering Foundations of Radio XML*, Wiley, 2006.

[18] E. Newcomer and G. Lomow, *Understanding SOA with Web Services*, Addison Wesley, 2005.

[19] M. Hightower and S. A. Pierce, The energy challenge, *Nature*, **452**, 285–286, 2008.

[20] International Energy Agency, *IEA World Energy Outlook 2005*, OECD/IEA, 2005.

[21] www.unesco.org/water/wwap/wwdr/wwdr1

2　On the possible developments for the structural materials relevant for future mobile devices

O. Ikkala and M. Heino

Materials science and processing technologies of electronics materials have been central to the rapid progress that we have experienced in electronics and its applications. Drastically new options are being realized and these can have a major impact on mobile devices, based on applications related to, e.g., flexible and polymer electronics and roll-to-roll processes. Mobile devices are, however, balanced combinations of electronics and peripheral structures, in which the latter are also strongly dependent on the advanced materials that are available. There is expected to be relevant progress in materials science contributing to the structural parts of mobile devices, e.g., promoting lightweight construction, rigidity, flexibility, adaptability, functional materials, sensing, dirt repellency, adhesive properties, and other surface properties. In this chapter we will review some of the potential future developments, leaving some others to be discussed in later chapters since they are intimately related to the developments of the electronics. We expect the most significant developments to take place in "soft" organic and polymeric materials and nanostructured organic–inorganic hybrids and therefore the main emphasis is there. Biological materials offer illustrative examples of advanced materials properties providing versatile lightweight, functional, and sustainable structures, showing potential for the design of novel materials.

2.1　Introduction

It is expected that new multifunctional materials with specific combinations of properties will extensively be developed for the structural parts and user interfaces of future mobile devices. Adaptivity, sensing, and context awareness are key sought-after features, and are described widely in the different chapters of this book. Rapid progress in materials science, in particular in nanostructured materials, will allow several new options for different customizations; references [1–6] describe the general concepts. In mobile devices, lightweight structural materials are essential, and it is expected that the evolving techniques will lead to materials that are tough and strong but also very flexible and resilient by incorporating nanocomposites [4] and bioinspired concepts

Nanotechnologies for Future Mobile Devices, eds. T. Ryhänen, M. A. Uusitalo, O. Ikkala, and A. Kärkkäinen. Published by Cambridge University Press. © Cambridge University Press 2010.

[6–8]. It would be exciting also to have self-healing materials developed to the bulk application stage [9–11]. Dirt repellent and self-cleaning materials [12–15] have already made an appearance in selected applications but it remains to be seen what their role will be in mobile devices. This chapter is mainly devoted to the concepts and potential applications of the new materials in the structural parts of mobile devices. Even though there is constant progress with metals (see, e.g., [4, 16]) the emphasis here is on polymeric and "bioinspired" organic and inorganic–organic materials as these might lead to more disruptive technologies. In addition, the new developments in organic and stretchable electronics and photonics [17–20] will lead to new options for designing the mobile device architecture, but as these aspects are closely related to signal processing, these will be discussed elsewhere in this book.

We will first discuss the current status of polymeric nanocomposites designed for use in lightweight structural parts where typically small amounts of nanoscale reinforcement are added mixed within polymer matrix [4, 6, 21–24]. Then we will present general scenarios in which self-assembly [1, 2, 25] is used to tailor material properties: after illustrating self-assembly with synthetic block copolymers [26, 27], we describe how nature has mastered a variety of functional materials and structures based on more complicated self-assemblies [8, 28, 29]. The biomimetic approach, i.e., understanding how nature has solved many material challenges and learning from these concepts, gives materials scientists a toolkit with which to design new lightweight and sustainable materials with attractive properties. This allows, e.g., a different approach to mechanically strong, fully organic or inorganic–organic nanocomposites. Finally, some specific functions such as self-healing [9–11], self-cleaning, and dirt repellency [12–15] are discussed. As materials science is developing very quickly, it is almost impossible to give a comprehensive overview of all relevant evolving materials properties related to delivering *Morph*-type of devices. Rather we have selected examples of how nanotechnology might be used to obtain useful materials properties. Ultimately, the critical issues with these new materials are cost, ease of manufacture, safety, lifetime, appearance, and sustainability aspects. Therefore, considerable uncertainties are involved in the prediction of future developments.

2.2   Nanocomposites by adding nanoscale reinforcements to tune the mechanical properties

Although many properties can be tuned using nanocomposite concepts, it is the mechanical properties that may be the most relevant in the present context. The challenge can be illustrated by the mechanical property charts [29, 30] that describe stiffness, as specified by Young's modulus, tensile strength, and toughness, where the last describes how well the material can suppress the catastrophic propagation of cracks, i.e., can suppress a brittle failure, see Figure 2.1. A variety of natural and man-made materials exist today with specific properties such as high strength, stiffness or elasticity, but good combinations of these properties are rare. Although no universal materials exist, arguments are given that clever molecular-scale tailoring could be a way to achieve synergistic properties. Of

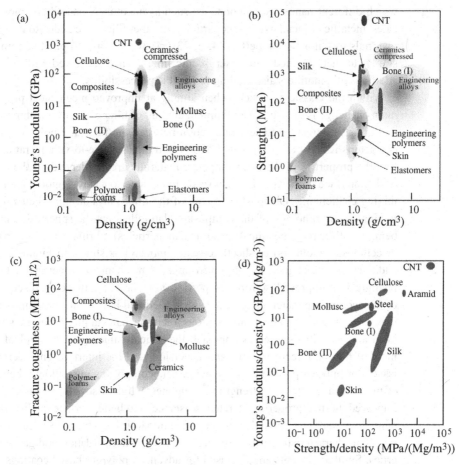

Figure 2.1 (a) Stiffness, (b) strength, (c) toughness as functions of density, and (d) specific stiffness vs. specific strength for some classes of materials. The figure illustrates that biological materials can have extraordinary mechanical properties at low density values. The figure has been adapted from [29].

course, there are also other relevant mechanical properties, such as fatigue strength and scratch resistance, but they are not discussed in the present general approach.

Nanotechnology and biomimetics bring new opportunities for the development of polymer-based materials [7, 8, 31–38]. In advanced applications it is essential to combine several, often conflicting, material properties. A prototypical example deals with efforts to combine high stiffness and strength with high toughness, where classically improvement of the first two properties leads to a deterioration in toughness, i.e. increased brittleness. On the other hand, brittleness can easily be suppressed, e.g., by selecting more rubberlike materials, but this is typically accompanied by reduced stiffness and strength. Therefore, multicomponent composites are routinely pursued in an effort to find practical optima [4, 6, 39]. A classic approach involves dispersing small amounts of reinforcing additives in a commodity polymer. This has led to an interest in nanocomposites in

which different nanoscale components, such as nanotubes, nanofibers, molecular silicates, metallic or metal oxide nanoparticles, or other fillers, are added to the commodity
materials to tailor the properties [4, 6, 33, 34, 36, 39, 40]. In practice reinforcement
similar to that obtained with macroscopic fillers can be achieved with much smaller
additions of nanofillers also leading to feasible processibility.

Within the classic approach, when aiming at improved mechanical properties, the
main challenge is to disperse a minimum amount of filler efficiently within the polymer
matrix and to create sufficient interaction between the filler and polymer matrix. The
characteristic feature of nanoscale fillers is their high surface-to-volume ratio, and therefore the properties of the nanocomposite materials greatly depend on the engineering
of the interface properties. But nanoscale fillers tend to easily agglomerate making the
mixing challenging. Obviously the type of the filler plays an important role in deter-
mining the morphology of the composite. Nanofillers can be spherical, and approach
being pointlike (e.g. metallic or ceramic nanoparticles), fibrillar (e.g. carbon nanotubes
or cellulose nanofibers) or sheetlike (e.g. nanoclay) as shown in Figure 2.2. Fibrillar
additives can be aligned for improved uniaxial mechanical properties or they can be
unaligned for improved isotropic properties. Note, however, that even if such composite
or nanocomposite materials are extensively used and have numerous improved properties, as shown by a wealth of applications, their mechanical properties are still far from
ideal and only in some cases approach theoretical values [41]. Control of the alignment of the reinforcement, efficient dispersion, and the interface engineering are key
issues for optimal properties. In general, nanocomposites may allow a better balance
of mechanical properties (strength vs. toughness), lighter structures, easier processing,
improved thermal properties, scratch resistance, and better surface quality. In addition,
nanofillers may give the polymers new functionalities, such as electrical conductivity,
magnetic characteristics, special optical effects, flame retardancy, and gas barrier properties. Similar concepts are also used for advanced polymer-based coatings and paints
for specific surface properties or structures.

2.2.1   Classic nanocomposites based on platelike reinforcements, as illustrated by nanoclay

As a particular example of inorganic platelike nanoscale fillers for nanocomposites, we
will discuss montmorillonites which are nanoclays [4, 39, 42–52]. Montmorillonites
are natural inorganic aluminum magnesium silicates of ca. 1 nm thickness with lateral
dimensions of ca. 100–1000 nm, which indicates a high aspect ratio. The predicted modulus is about 270 GPa [53]. Normally nanoclays are negatively charged in comparison
to the silicate layers due to an isomorphic substitution of Si^{4+} with Al^{3+} or Mg^{2+}. The
negative charge of the layers is compensated by positive counterions. Over the years,
there have been extensive efforts to add a minor concentration of nanoclay to different
engineering polymers in order to improve the mechanical properties. Nanoclays tend to
aggregate and to form stacks. The individual nanoclay sheets within the polymer matrix
(Figure 2.2(c)) need therefore to be exfoliated. The exfoliation can be achieved by
applying positively charged surfactants or polymers that bind to the negatively charged

Dimensionality of the reinforcing fillers in nanocomposites

Sheetlike (two-dimensional)

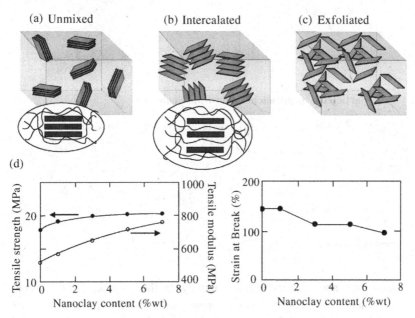

Figure 2.2 Schemes for "classic" polymeric nanocomposites with small concentrations of the nanofillers, aimed at improving material properties in comparison to the pure matrix polymer. The top part illustrates the different states of aggregation of the layered (two-dimensional) sheetlike nanofillers: (a) the unmixed state where the nanofillers macroscopically phase separate from the matrix polymer and give little improvement of the mechanical properties; (b) the intercalated state, where the matrix penetrates the stack of the nanosheets; and (c) the exfoliated state, where the nanosheets have been well dispersed within the matrix to allow a considerable improvement of the properties. (d) From the very extensive literature [39], an example has been selected showing typical mechanical properties for polyethylene nanocomposite where surface functionalized nanoclay has been added, adapted from [54]. It illustrates that most typically the stiffness and strength have been improved at the cost of ductility. (e) Long and entangled nanofibers reinforcing the matrix. The selected examples are surface functionalized carbon nanotubes or cellulose nanofibers. (f) Short nanorods can also reinforce the matrix, but there the engineering of interface bonding between the fibers and the matrix can become particularly critical. The whisker-like cellulosic nanorods are one example. [55, 56] (g) Nanoparticles can also be used to tune the properties, as they modify matrix polymer behavior near the nanoparticle surfaces [4, 22]. Such nanocomposites have been used, e.g., to control the crack propagation properties for increased toughness. Considerable effort has been given to obtaining higher levels of control of nanoparticle assemblies, in particular using block copolymers to confine the nanoparticles within the interfaces or nanodomains of self-assemblies ((h), (i)). Such materials have been aimed at e.g., optical applications.

Fiberlike (one-dimensional)

(e) Long percolating and entangled nanofibers   (f) Short nanorods

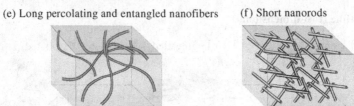

Nanoparticlelike (approaching zero-dimensional)

(g) Random   (h) Interface bound   (i) Confined

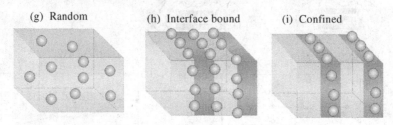

Figure 2.2 (*cont.*)

nanoclays and mediate surface activity to the polymer matrix. Typically a small percentage of nanoclay gives improved strength and stiffness but characteristically toughness decreases. This is a typical problem in polymer reinforcing but careful optimization allows practical combinations of properties for several applications. Later we will describe a biomimetic approach which can be used to obtain even better properties.

2.2.2 Nanocomposites based on fibrillar reinforcements, as illustrated by carbon nanotubes (CNTs) and cellulose nanofibrils

Nanocomposites consisting of fibrous reinforcements constitute another major classic example, see, e.g., [4, 51, 57]. Carbon fibers have long been used to reinforce polymers for various bulk applications but interest has shifted towards CNTs [34, 37, 40, 41, 58–74]. The single wall CNT has the most spectacular properties, but due to cost optimization, multiwall CNTs and even less pure materials have also been used. The combination of high aspect ratio, small lateral size, low density, and more importantly, excellent physical properties such as extremely high tensile strength (in the range of 30–60 GPa), high tensile modulus (ca. 1000 GPa), and high electrical and thermal conductivity make single wall CNTs attractive candidates for reinforcing fillers in high strength and lightweight multifunctional polymer nanocomposites.

Chemical functionalization of nanotubes is the key to enabling good interaction between CNTs and the polymer matrix and to guaranteeing efficient load transfer under mechanical stresses (see Figure 2.3(b)–(d)). In situ polymerization is an effective way to create composites with good dispersion of CNTs, well bonded to the polymer. Controlling the interactions at the CNT–polymer interface allows lightweight polymer composites with tailored properties to be manufactured. For electrical conductivity a continuous

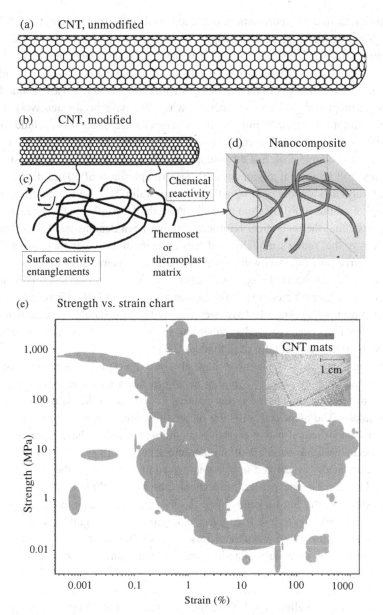

Figure 2.3 (a) A single wall CNT and (b) a schematic showing its chemical modifications. These modifications promote interaction with the polymer matrix by (c) direct chemical links and chain entanglements to reinforce (d) the nanocomposites. (e) To illustrate the potential of CNTs for mechanically strong materials, (e) the strain and strength of CNT interwoven mats (see inset) are compared with typical values for commodity materials as represented by the gray area. In part reproduced with permission from [62].

network structure, i.e., percolation, is needed, while uniform alignment of CNTs is beneficial for high uniaxial tensile strength. Thermoset polymers have been widely used as the matrix for CNT composites, and have been used in commercial products such as windmill blades, golf clubs, tennis rackets, ice-hockey sticks, skis, etc. The addition of chemically modified CNTs increases the fracture toughness (typically about 40–60%). Also thermoplastic moldable composites with CNT have been extensively developed. Some examples of reported improvements in mechanical properties include an increase of stiffness by 214% and strength by 162% at a multiwall CNT loading of 2 wt% in polyamide-6 composites, made by simple melt-mixing and then compression molded to films with no special alignment [66]. Supertough fibers of 50 μm diameter consisting of polyvinyl alcohol with 60 wt% of single wall CNTs have been produced with a Young's modulus of 80 GPa and a tensile strength of 1.8 GPa [75]. Significant progress in approaching the theoretical reinforcement has been demonstrated [41]. Normalized for density the Young's modulus and tensile strength are more than twice the values for the corresponding steel wire. For polyamide-6, 10 composites incorporating 1 wt% chemically functionalized single wall CNTs, and so-called interfacial polymerization, 160% higher Young's modulus, 160% higher strength, and 140% higher toughness compared to neat polyamide-6,10 have been reported [76]. It should be noted that electrical conductivity has been reported at rather low amounts of CNTs, which is due to the high conductivity and aspect ratio of individual CNTs.

Several other types of nanofibers have also been investigated as reinforcing additives for commodity polymers [4]. Another nanofibril that could have substantial impact in future nanocomposites as a "green material" is based on the native crystalline form of cellulose. The plant cell walls are hierarchical multicomponent materials in which the smallest structural native crystalline cellulose nanofibers have a diameter of a few nanometers and they form 10–30 nm nanofiber bundles [8, 56, 77, 78]. These, in turn, are hierarchically assembled into larger structural units within the less ordered background matrix. The mechanical properties of the native cellulose nanofibers are not yet completely known, but they are suggested to have a very high tensile modulus, up to ca. 140 GPa, and the tensile strength is expected to be very good, up to the GPa range [79]. Such values are not as high as those of CNT, but the material is attractive due to its sustainability, its wide availability, and its facile functionalization. However, upon dissolution, the favorable native crystal structure is lost and in order to preserve it, the nanofibers have to be cleaved from the macroscopic cellulose fibers without dissolution. This has proved challenging, but the methods required to do this are being developed (see Figure 2.4): processes have been identified to disintegrate the disordered material both between the nanofibers and within the nanofibers to cleave highly crystalline native cellulose rodlike whiskers with a diameter of a few nanometers [77]. Due to their rodlike character, they form liquid crystal solutions. They have been mixed with polymer matrixes and give considerable reinforcement and even "switchable" stiffening [55, 56, 80]. On the other hand, Figure 2.4 also illustrates that selective disintegration of only the interfibrillar disordered material allows long and entangled native cellulose nanofibers of a nanometer-scale diameter [81–83]. Such nanofibers have been used to make strong "nanopapers" [84] and to prepare nanocomposites. To illustrate

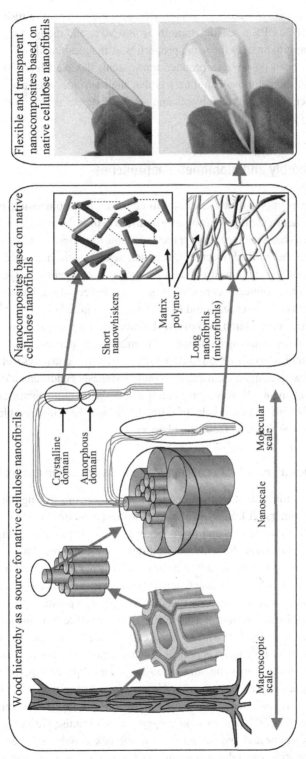

Figure 2.4 The hierarchical structure of plant cell walls allows them to be disintegrated, providing native cellulose nanofibers with good mechanical properties with crystalline and amorphous domains. The figure shows that there are established ways of achieving either short rodlike nanofibrils (by cleaving the intra- and inter-fibrillar amorphous material) or long and entangled nanofibrils (by cleaving only interfibrillar amorphous material) by modifying the preparation processes. The nanocomposites can be both transparent and flexible [86]. They can allow "green" sustainable reinforcement in nanocomposites within commodity polymers. In part modified by permission from [86, 87].

the latter, cellulose nanofibril composites with cured phenolic resins can have tensile modulus of 20 GPa, a tensile strength of 400 MPa, and elongation of several percent, [85], and approximately the same properties as magnesium alloys, but lower densities. It has also been possible to produce transparent, flexible films [79]. Finally, a useful property of the cellulose nanofibers is that their surfaces are easy to modify to allow good dispersibility within solvents and several types of matrix polymers and to allow bonding to various matrices.

2.3   Self-assembly and biomimetic approaches

The above examples describe efforts to construct nanocomposites by mixing a minor amount of nanoscale fillers in polymer matrices to tune the material's properties. Such an approach has been in many ways "practical": how to achieve a sufficiently large improvement in the mechanical, electrical, thermal, scratch resistance, or diffusion properties with the smallest amount of the additive to give a competitive edge in products. Even though this has resulted in considerable success, one could ask whether better synergistic combinations of properties could be achieved, as augured by Figure 2.1. In this section, we will turn our attention to biological materials and the underlying concepts therein; they can provide amazing combinations of material properties using self-assemblies and their hierarchies based on biological polymers, proteins, and polysaccharides, sometimes combined with inorganics, using only mild synthesis conditions and only a few types of atomic elements. However, as natural materials are quite complicated to understand from the first principles, it is helpful first to discuss the self-assembled materials based on the simplest forms of block copolymers.

2.3.1   Block copolymers: a facile concept for self-assembly and applications

We will first illustrate the general concept of self-assembly using the simplest forms of block copolymers [26]. Self-assembled nanostructures form due to competing attractive and repulsive interactions and this concept is illustrated by the examples in Figure 2.5, which also illustrates the wide possibilities to tune the nanostructures. The different structures, in turn, lead to different properties, allowing new applications, as will be discussed later. A characteristic feature of polymers is that mixing two or more different polymers almost always leads to macroscopic phase separation, i.e., different polymers are immiscible and repulsive. This means that in polymer mixtures, the minor polymer component typically forms macroscopically phase separated "droplet-like" domains within the major polymer component. Such structures are poorly controlled and highly dependent on the processing, and in most cases have poor properties. But the situation changes when different types of polymer chains are mutually chemically connected in order to prevent their macroscopic phase separation, thus constructing e.g., block copolymers [26, 27]. For example, suppose two immiscible polymers A and B are chemically connected end-to-end to form diblock copolymers, see Figure 2.5(a)–(g). Phase separation can take place only as far as the polymers can stretch, i.e., at the

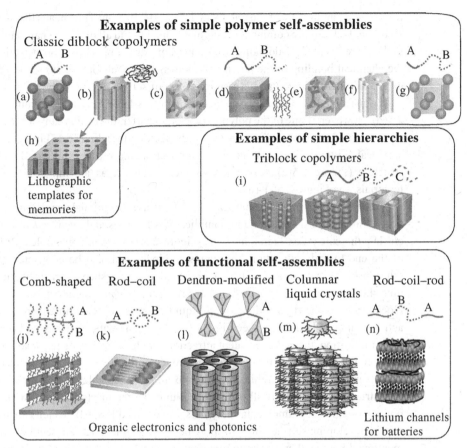

Figure 2.5 Examples of self-assemblies to illustrate the power of constructing a great variety of nanostructures in a spontaneous manner. Two mutually repulsive polymers A and B covalently connected end-to-end lead to lamellar (d), network-like so-called double gyroid (c), (e), cylindrical (b), (f), or spherical (a), (g) self-assemblies. The concept (h) has been extensively developed for nanoscale templates for memories [92–94]. Adding a third repulsive block C allows further structures and constitutes the simplest case for structures at different length scales, i.e., hierarchies (i). Also rodlike liquid crystalline [95] or conjugated electroactive chains [18] have been incorporated within the self-assemblies to allow functionalities. The comb-shaped architecture (j) incorporates side chains, e.g., to control the separation of the electroactive polymer chains for soft electronic materials [18, 96]. Another concept (k) for electroactive assemblies is based on fibrillar constructs of conjugated rods equipped with coillike end-groups [97]. Concept (l) demonstrates that detailed control of the side chain packing by wedge-shaped side chains can lead to detailed control of the self-assemblies, to allow e.g., columnar packing [98]. A useful concept for columnar packing is illustrated by concept (m) based on disk-like chemical groups, which can be electroactive to allow electrically transporting columns for molecular electronics [18, 99]. Concept (n) has been discussed in the context of providing conducting lithium nanochannels for batteries [100].

nanoscopic length scale of the polymers, i.e. ca. 10–100 nm. This means that in this case, all of the A-chains are confined in their own domains and all of the B-chains are confined in the corresponding B-domains. Such a periodic structure is based on attraction (due to the chemical bonding of the polymeric blocks) and repulsion (due to the immiscibility of the polymeric blocks) and is a simple form of self-assembly. If the A- and B-chains have equal lengths, the self-assembled structure is lamellar, due to symmetry, Figure 2.5(d). However, if one of the polymer blocks is shorter, the chains cannot form lamellae with an equal density, and curvature results, leading to cylindrical (Figure 2.5(b), (f)), spherical (Figure 2.5(a), (g)) structures, as well as more a networklike double gyroid structure (Figure 2.5(c), (e)). Such structures can act as a template for functional materials, see Figure 2.5(j)–(n).

Naturally, increasing the number of blocks causes additional repulsions and attractions, which in turn lead to more complicated structures and even structures at different length scales, i.e., hierarchies, see Figure 2.5(i) for some examples [26]. Instead of the end-to-end connection, the repulsive blocks can also be connected in various other fashions, e.g., in a star-shaped, branched, or comb-shaped brush-like manner, to tune the structures, see Figure 2.5(j) for the comb-shaped architecture. The polymer blocks can also be rigid rodlike, as in liquid crystals and even in electro- and photo-active conjugated polymers (see Figure 2.5(k), (n)), as will be discussed briefly below [88]. In addition, instead of covalent attractions, sufficiently strong physical interactions between the repulsive groups (such as hydrogen and ionic bonds) can be used to drive self-assemblies and hierarchies, in order to allow switchable electro- or photo-active nanostructures [89, 90]. If all of the repeat units are connected only by weak physical bonds, so-called supramolecular polymers are obtained [88, 91], which can also have a role in developing self-healing [10] polymers, see Figure 2.12. These all warrant being mentioned here as they illustrate the great potential to perform nanoscale engineering which allows in-depth modification of many properties, which is central for the so-called soft nanotechnology.

Selected applications of block copolymeric self-assembled structures are next reviewed to demonstrate the power of the concept. The classic application of block copolymers is in thermoplastic elastomers [101]. Historically, elastomers, i.e., rubbers, have been achieved by chemically cross-linking polymer chains of natural rubber to make networks, where the network architecture is relevant for elastic recovery of the shape after stretching. But when cross-linked, the material cannot be reprocessed due to the permanently formed covalent cross-links. Block copolymers allow a classic alternative route. Consider a block copolymer consisting of three polymer blocks with a soft middle block B and hard glassy end-blocks A, which form spherical self-assemblies with the matrix of the soft block. The self-assembled glassy end-block domains act as cross-links for the soft middle blocks at room temperature as they are hard and glassy but allow processing above the glass transition temperature. Such concepts allow melt-processable thermoplastic elastomers which are widely used as rubberlike materials in bulk applications such as in shoes, tyres, asphalt, to name a few. Therefore thermoplastic rubbers can be regarded as nanocomposites of glassy and soft materials, and demonstrate that nanocomposites can have very different properties from each of the constituents. They also demonstrate that self-assembled structures are not just a scientific curiosity but can have widespread bulk applications.

As well as classic bulk applications, block copolymers have more recently aroused expectations related to more advanced applications in which the self-assembled periodic nanostructures are specifically used for nanoscale construction. The example of block copolymer lithography is shown in Figure 2.5(h) [92]. In one approach, polymethylmethacrylate, a known ultraviolet-sensitive lithographic resist polymer, is selected as block A in a block copolymer, in which the other block, block B, is a ultraviolet-inert polymer, such as polystyrene. Concepts have been developed to allow cylindrical self-assembly (Figure 2.5(b)) in thin films in which all the polymethylmethacrylate nanoscale cylinders are standing upright on a substrate (see Figure 2.5(h)). Upon applying ultraviolet light, one can degrade the polymethylmethacrylate material within the cylinders and only the polystyrene matrix remains. Therefore a dense array of emptied nanometer-length-scale cylindrical holes is created which can, e.g., be subsequently filled with magnetic material by evaporation to produce a magnetic memory material. Related structures [93] for magnetic materials have been suggested that give a nominal 1 terabyte/inch^2 memory capacity [94]. This concept shows the power of self-assembly as a template for preparing nanostructured materials utilizing various forms of block copolymer lithography to allow patterning of substrates with different functional units. Another example is the use of block copolymers in solar cells: the double gyroid morphology (Figure 2.5(c), (e)) is used to prepare nanostructured Grätzel-type [102, 103] dye-sensitized solar cells [104]. Therein, the interconnected network phase of the gyroidal self-assembly is constructed using a degradable polymer block, and subsequently the network is emptied. Titanium dioxide can then be deposited in the open channel. Even though the performance has not yet surpassed other related preparation schemes, it is expected that the nanoscale structural control might pave the way for future controlled self-assemblies for solar energy harvesting.

An example demonstrating developments for functional self-assemblies is found in block copolymers with electroactive polymer [6, 18, 97] blocks. So-called conjugated polymers consist of alternating single and double carbon–carbon bonds that lead to semiconductivity or even conductivity upon doping, and therefore to electro- and photoactive polymers, see Figure 2.6 for a few examples. Due to this structure, these polymer chains adopt a totally rigid rodlike or semirigid shape. One can even say that CNTs belong to this group of materials. Due to the rigidity, the conjugated chains are prone to aggregation and it is not surprising that in the general case they do not melt and they can be poorly soluble in common solvents. This complicates the processability, which is particularly detrimental as one of the main reasons for using polymers is to incorporate facile polymer processing techniques, such as spin-coating, molding, etc. However, the aggregation can be reduced and easy processability achieved by incorporating appropriate side chains (Figure 2.5(j)) or end-chains (Figure 2.5(k)) in the conjugated backbone, see Figure 2.6. The combination of the rigid electroactive backbone and the side chains lead to self-assemblies, as shown schematically in Figure 2.5(j), (k), where the semiconducting polymer backbones are separated on the nanometer length scale due to the self-assembly caused by the side chains. An example of a relevant

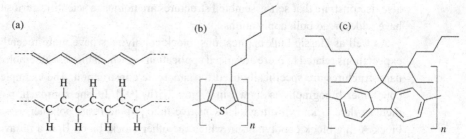

Figure 2.6 Examples of conjugated polymers. (a) The simplest prototypic material is polyacetylene, which illustrates the alternating single and double carbon bonds. (b) Poly(hexylthiophene) and (c) poly(diethylhexylfluorene) are examples of comb-shaped self-assembling semiconducting polymers, of which (c) in particular may be relevant for applications. Poly(hexylthiophene) self-assembles as stacked layers (Figure 2.5(j)), whereas poly(diethylhexylfluorene) self-assembles in a hexagonal cylindrical manner (Figure 2.5(m) shows this a type of symmetry).

material class is side chain modified polyfluorenes, such as shown in Figure 2.6(c), which self-assemble to form soluble and processable material with desirable properties. This material is expected to have commercial potential in soft electronics and photonics. A commercial electrically conducting polymer suitable for soft electronics is based on modified polythiophenes and their complexes with an acidic polymer as well as polyaniline. Such materials allow conductivity levels of up to a few tens of siemens per centimeter.

2.3.2   Biological materials: selected examples

The above examples demonstrated structure formation based on the simplest forms of self-assembly and some of its applications. Next we discuss a few aspects of biological materials from a materials science point of view. We will first briefly describe why the hierarchical self-assemblies and nanocomposite structures of biological materials allow feasible synergistic combinations of properties [8, 28, 29].

It is helpful first to clarify some definitions. (1) Biological materials are encountered only in the natural organisms and are synthesized within the biological processes. Their structures and functions have evolved under the different development pressures. (2) Bioinspired materials are "man-made" materials synthesized in laboratories in which selected biological concepts are imitated, with the aim of obtaining new types of structures and functionalities. (3) Biomaterials e.g., man-made implants are materials that are intended to "work" under biological conditions.

From the materials science point of view, biological materials have structures, mechanical properties, and sometimes even other functional properties that are unique and warrant special attention. They vary widely: typically their density, $\rho < 3 \, \text{g/cm}^3$, is low in comparison with most man-made structural materials. Their tensile moduli (stiffness) are in the range of 0.0001–100 GPa and tensile strength in the range of 0.1–1000 MPa, i.e. they range from very soft gellike materials to almost steellike materials. Obviously, evolution has favored material concepts in which light-weight construction is combined with good mechanical properties. This is the reason why they can be of a special interest within the scope of the present book to promote portability and wearability.

Before considering selected examples, there are general features that warrant comparison with synthetic materials. In biological materials only few atomic elements are used, the (bio)synthesis conditions are mild, the structures are hierarchical with evolution resulting in structures at the length scale relevant to a given property, the materials are adaptive and multifunctional, and they can even self-heal after a damage. Their structures and the related properties are the result of a lengthy evolution process. Still, there are some common features; the materials are almost invariably nanocomposites, having reinforcing domains to promote stiffness and strength and plasticizing dissipative domains to suppress brittleness. This suggests some attention should be paid to the generic concepts within the biological materials. Next we will discuss two almost classic examples of biological nanocomposite materials in which hard and soft domains are combined in order to obtain synergistic mechanical properties: silks and nacreous abalone.

2.3.2.1 Example of a totally organic nanocomposite biological material with feasible mechanical properties: silk

Silk is a proteinic polymer which has become a classic example of biological materials due to its extraordinary mechanical properties which are based on an all-organic nanocomposite structure [8, 105–111]. It has been, and still remains, a considerable source of inspiration for materials scientists. In fact, silk is not a single material, but exists as several variants. Silks consist of proteins, which are polymer chains that have well-defined sequences of amino acid repeating units. Each repeating unit of a protein contains peptide groups, $-NHCO-$, and altogether there are 20 variants of these units as each of them also has a characteristic side group. At this general level it suffices to mention that some of the repeating units are hydrophilic (attract water), some of them are hydrophobic (repel water), and some of them contain specific interacting groups. For each type of protein there is a specific sequence of the 20 different repeating units leading to a specific self-assembled shape of the polymer chain and in turn higher-order structures, which is the basis for the biological functionality. The structure is formed by self-assembly which we have already encountered in a simpler form: remembering that simple block copolymers form self-assembled periodic structures due to the competition of attractions and repulsions between the polymeric blocks, one could say that the protein folds to a specific conformation and forms self-assembled structures based on the same logic, i.e., as specified by the attractions and repulsions and steric requirements of the amino acid sequences. But in this case, the design rules for the self-assemblies, as developed during their lengthy evolution, are very much more complicated than in the simple block copolymers.

For conciseness here we limit ourselves to considering spider silks. A spider's web consists of three different types of silk, see Figure 2.7: (1) drag-line silk, which supports the web in the external environment; (2) web-frame silk which is used to construct the radial parts; and (3) viscid silk which forms the spiral parts. Drag-line silk has very high strength, see Figure 2.7. Its tensile modulus is in the region of 10 GPa, its tensile strength

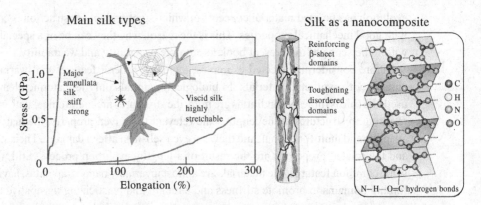

Figure 2.7 Different forms of silk and their mechanical behavior are illustrated. The hard and aligned protein nanocrystals consist of stretched protein chains which pack efficiently to give reinforcement to promote stiffness and strength. The matrix consists of soft protein chains, promoting toughness. Modified based on [107].

is about 1 GPa and its tensile strain is about 6%, i.e., the material is stiff, strong, but still stretchable. Silk compares favorably with steel, but has a much lower density. Therefore, amazingly spiders can prepare a nearly steellike material under mild conditions without metallic ingredients. This classic observation has a broad significance for all fields of polymer science, suggesting that, in principle, polymer materials should still have tremendous development potential compared to the present state-of-the-art engineering polymers. Viscid silk has a different function, as it needs to be highly extensible to dissipate the kinetic energy of the caught prey and sticky enough to bind that prey. Such fibers have a maximum strain of 2000–3000%, which is as much as the best synthetic rubbers.

The drag-line silk structure is discussed here as an example of the general design principles needed to optimize mechanical properties. The structure is a fully organic aligned nanocomposite consisting of a high loading of crystalline nanodomains in a flexible matrix. The crystalline domains are formed by antiparallel and stretched proteinic chains that pack very efficiently due to the high number of interchain hydrogen bonds to form well-defined nanosized crystals. Because of the stretched chains, and their mutual bonds, the nanocrystals are strong and are able to act as reinforcement to promote strength and stiffness, see Figure 2.7. On the other hand, the amorphous matrix of silk consists of various helical and coillike chains, and such conformations promote "rubberlike" behavior. The matrix also contains water which further plasticizes it, thus contributing to the elastic and dissipative characteristics needed for toughness. Therefore, silk is a nanocomposite consisting of hard reinforcing nanocrystals within a soft material, where the balance between these competitive properties is such that its allows a high modulus, high strength, and high toughness. Very interestingly, although all of the starting materials are water-soluble, the fiber spinning process leads to water-insoluble fibers. Furthermore, the spinning process leads to the alignment of the reinforcing crystals in parallel to the fiber promoting the mechanical properties, see Figure 2.7. Even though various different analogous "semicrystalline" polymers, with crystalline reinforcing

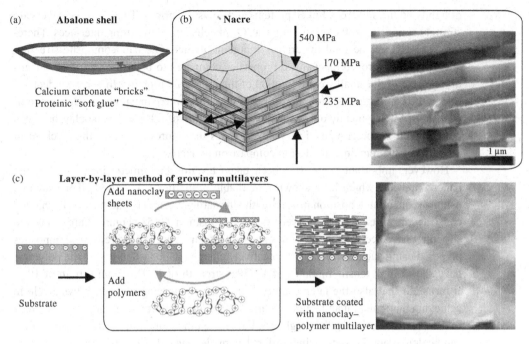

Figure 2.8 (a) The nacreous layer of mollusk shell is an example of a biological organic–inorganic nanocomposite; (b) it shows a synergistic combination of high stiffness, high strength, and high toughness (electron microscopy image: the layered "brickwall" structure). (c) A biomimetic approach for nacrelike material using layer-by-layer deposition of alternating layers of negatively charged nanoclays and positively charged polymers and the "pearly" visual image due to the layered structure. Modified with permission, based on [8, 31].

nanodomains and amorphous energy absorbing domains, have been synthesized it has remained difficult to achieve properties comparable to those of silk, especially if economic starting materials are used. But, "silk-mimetic" materials may ultimately be developed, in which the essential design principles are combined with production friendly materials.

2.3.2.2 Example of a organic–inorganic nanocomposite: nacreous shell

In nature there are numerous inorganic–organic ordered hierarchical structures that have outstanding mechanical properties in comparison to their density, such as the nacreous layer of mollusk shell, bones, and teeth [8, 112–114]. These all combine light-weight structures with extraordinary mechanical properties. As a specific example, here we consider the nacreous layer of mollusk shell which has remarkable mechanical properties in comparison with most man-made materials: its tensile modulus is ca. 64 GPa and its tensile strength is ca. 130 MPa, i.e., the material is very stiff and strong. However, in addition it is tough, showing reduced brittleness, unlike most man-made reinforced materials. Nacre is a nanocomposite material, consisting of thin, 0.3–0.5 µm, calcium carbonate ($CaCO_3$) sheets which act as reinforcement to promote stiffness and strength (Figure 2.8). In between these sheets, there are thin, 20–30 nm polymer layers, consisting of tightly bound and cross-linked proteins and polysaccharides. These "softer" layers are strongly connected to the reinforcing $CaCO_3$ platelets to allow strong interfaces. Therefore, ca. 95% of the total material is inorganic matter. The nanocomposite structure allows much improved mechanical properties compared to bulk $CaCO_3$, which means that in spite of the small weight fraction of the interlaced proteins, they have a decisive role in the compositions. Note that nacre is therefore unlike most man-made nanocomposites in which typically only a small weight fraction of filler, e.g., nanoclay, has been incorporated in the composition. In these cases the improvements of the mechanical properties have remained modest in comparison to nacre.

However, important steps have been taken towards biomimetic inorganic–organic nanocomposites which have shown remarkable properties [31, 35, 38]. For example, using a sequential deposition process with alternating negatively and positively charged construction units, so-called layer-by-layer deposition [115] (Figure 2.8(c)), layered structures have been obtained that in some respect mimic nacre. Layered nanocomposites prepared using such concepts with negatively charged nanoclay and synthetic polymers have shown high tensile modulus of 13 GPa, strength of 109 MPa, and strain of 10%. It turns out that also the lateral dimension of the reinforcing platelets plays a role in achieving flaw-tolerant and tough materials. If the platelets are too large, they can break in deformation, reducing the toughness, but on the other hand if they are too small, they can be drawn out. By optimizing, differently made, layered structures strengths as high as 300 MPa and large elongations of 25%, have been achieved.

2.3.2.3   DNA-origami templating: towards aperiodic self-assemblies

Self-assembled structures based on block copolymers have led to periodic or quasiperiodic structures. Even though several interesting properties can be achieved in periodic nanostructures, most engineering structures are not periodic. If bottom-up construction based on self-assembly is to become a general manufacturing method, models for building nonperiodic structures based on self-assembly must be developed. We have already seen that at the length scale of individual proteins, nonperiodic self-assembly leads to folding of proteins into the characteristic polymeric shapes or conformations. A general concept for nonperiodic self-assembly for bottom-up construction is based on DNA, see Figure 2.9 [116–119]. This will not be ready for industrial applications in the near future, but conceptually it is interesting and it may open new avenues for material engineering at the nanoscale in the longer term. We emphasize that other concepts allowing aperiodic "programmed" self-assemblies are at present very limited.

The concept is based on the double-strand pairing of DNA in which each of the four bases, i.e. adenosine (A), thymine (T), cytosine (C), and guanine (G), invariably has its one-to-one matching counterpart base forming the base pairs AT and CG. This is the foundation on which the biosynthesis of proteins in biological organisms is built. Now, however, it has been realized that such pairing can be used to code nonbiological structures as well [116–118]. The logic is, in principle, simple: in biological coding, the AT and CG pairs follow each other smoothly and a long DNA double helix is formed. Within the double helix, one could regard one helix turn as a pixel of size 2 nm × 3.6 nm (Figure 2.9(c)). In one artificial construct, often called DNA-origami, one uses a long

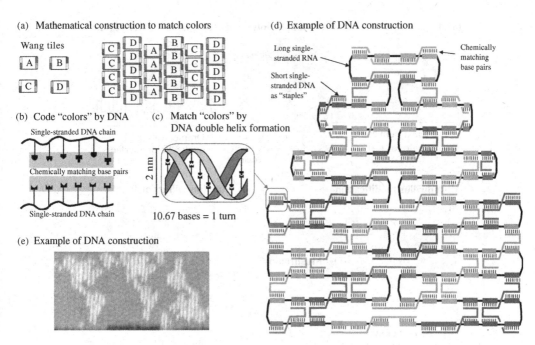

(a) Mathematical construction to match colors

(d) Example of DNA construction

(b) Code "colors" by DNA

(c) Match "colors" by DNA double helix formation

(e) Example of DNA construction

Figure 2.9 The concept of DNA-origami allows programmed assembly based on the "pixel size" 2 nm × 3.6 nm. Reproduced and redrawn in part with permission from [117, 119].

single-stranded nucleic acid, which is artificially designed to be "folded" according to a predetermined pattern, see Figure 2.9(d) (the black solid line). To define the spatial separation of the nearby folds, the above pixel sizes must be followed. Then, short pieces of single-stranded DNA are slotted between the long folded nucleic acid chain, where the sequences of the AT and CG base pairs have to be designed to allow the short single-stranded DNA chain to recognize their proper loci. In this way designing single-stranded nucleic acid chains with desired sequences of A, T, C, and G becomes a mathematical problem. Suitable algorithms have now become available and various prototypical shapes have been programmed. It is too early to predict the importance of the concept for future materials science but conceptually building "programmed" materials using DNA-origamis for bottom-up aperiodic construction [120–122] holds much promise.

2.4 Functional surfaces

2.4.1 Controlled surface properties and dirt repellency

The wetting properties of surfaces have been extensively studied [12–15, 123–127]. Dirt repellency and self-cleaning of the surfaces could be interesting features in future mobile devices. Here we discuss specific wetting properties and dirt repellency as other forms of bioinspiration. Over the last 10–15 years a lot of effort has gone into designing surfaces that do not allow wetting by water or other liquids, therefore leading

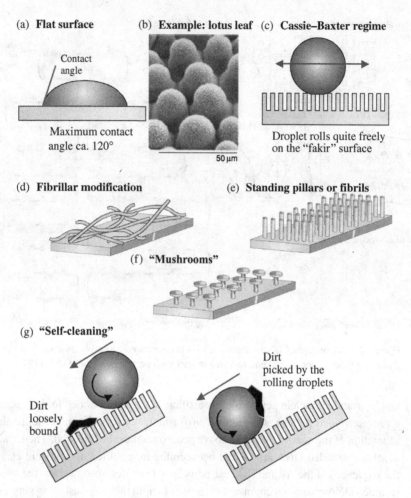

(a) **Flat surface**   (b) **Example: lotus leaf** (c) **Cassie–Baxter regime**

Contact
angle

Maximum contact
angle ca. 120°

50 μm

Droplet rolls quite freely
on the "fakir" surface

(d) **Fibrillar modification**   (e) **Standing pillars or fibrils**

(f) **"Mushrooms"**

(g) **"Self-cleaning"**

Dirt
picked by the
rolling droplets

Dirt
loosely
bound

Figure 2.10 Routes to tailored surfaces that repel liquids and dirt. (a) A droplet of liquid (e.g.,
water) on a surface, showing the contact angle. For water the contact angle remains below ca.
120° for typical flat surfaces. (b) A lotus leaf is a biological example of a superhydrophobic and
self-cleaning surface (reproduced with permission from [129]). (c) For low-energy surface
modification and when the surface topography is designed to allow air pockets to remain, the
water droplet hangs on "fakir-mat-like" pillars, the contact angle becomes very high and the
droplet can easily roll over the surface. Surface roughness can be increased by various ways: (d)
nanofibrillar constructs, e.g., by electrospinning or etching; (e) vertical pillars or fibers; (f)
mushroom-shaped pillars that allow particularly effective superhydrophobility [14]. (g)
Schematics of dirt repellency and self-cleaning [125].

to large contact angles (see Figure 2.10(a)), which can lead to dirt repellency. This work
was inspired by the leaves of the lotus plant [128], on which water droplets tend to adopt
an almost spherical shape, i.e. the contact angle is >150°. The wetting properties of various liquids on surfaces have been investigated, and there exists a classical relationship
between the contact angle and the surface energies between the surface and the air, the
surface and the liquid, and the liquid and the air [12]. In simple terms, oillike (nonpolar)

liquids do not like to wet polar surfaces and lead to a high contact angle, and conversely water (polar) does not like to wet oil-like surfaces, which are called hydrophobic. To reduce the wetting tendency of water on substrates, i.e., to increase the contact angle, it is important to select materials that have low surface energy towards air. Fluorination provides the most efficient surface coating with the maximum contact angle typically limited to ca. 120° for a flat surface. This has a wealth of technological applications. All other surface coatings for flat surfaces have a smaller contact angle. However, the lotus plant with its superhydrophobic leaves has a contact angle >150° without its surface being fluorinated which has aroused considerable interest. The lotus's super-hydrophobicity is due to the chemical structure of the surface combined with its hierar-chical surface topography, see Figure 2.10(b). Electron micrographs show structures at the tens of micrometer scale and also at the nanoscale, and it is thought that the combined effect is central for the wetting properties. Once this was realized, numerous routes were taken to construct surface patterns to control the wetting. Surface patterns can lead to the so-called Cassie–Baxter regime (Figure 2.10(c)) leading to superhydrophobicity, in which air remains trapped within the surface topography and the water droplet is there-fore suspended on the "fakir" patterns; this can even allow quite free rolling of water droplets. On the other hand, there are variations designed to introduce dirt repellency. One of these also relies on the Cassie–Baxter regime. In this, if the surface energy of the substrate is very small, still incorporating the surface roughness, dirt may not adhere on the surface and the rolling water droplets can sweep the dirt away.

Various ways of achieving superhydrophobic surfaces with water contact angles >150°, and even dirt repellency in some cases, have been reported. Electrospinning of fibers allows 100–200 nm fibers to achieve rough surface topographies, soft and optical lithographic patterning has been used, and various fiber formation protocols have been employed to produce essentially perpendicular "forests" of nanofibers to achieve "fakir-mat-like" surfaces. For metals one can achieve suitable surface patterns and surface chemical functionalizations using etching techniques. One effort resulted in "overhang" structures, such as the lithographically prepared "mushroom-shaped" pillars, see Figure 2.10(f) [14]. In this the Cassie–Baxter regime is easily fulfilled as the liquid droplets, due to their surface energies, cannot easily deform to wet the surface patterns. Such coatings have also demonstrated the possibility of superoleophobic surfaces, which lead to very high contact angles for oils [14]. In summary, the necessary conditions for super-hydrophobic and superoleophobic surfaces are starting to be understood at the scientific level, but producing the surface in a production friendly manner on the large scale, with high wear resistance, still remains a challenge. We expect wetting phenomena to become important in everyday devices once feasible and scalable production technologies have been developed and the wear resistance properties of the nanostructures are settled.

2.4.2   Surface-adhesion-based bioinspired gecko effects

For fully adjustable and deformable mobile devices as suggested by the Morph concept, one could visualize options in which the device can be suspended on various surfaces using adhesive forces. There are many mechanisms for adhesion and typically they

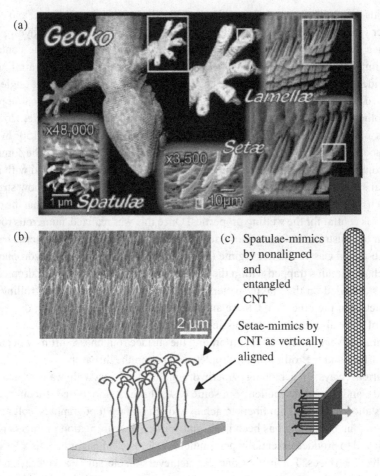

Figure 2.11 Adhesion inspired by biological systems: (a) the gecko and its hierarchically structured foot pad structure which allow it to adhere to a vertical surface; (b), (c) multiwall CNT forest used as a gecko-inspired adhesive (partly reproduced by permission from [137, 138]); part (a) courtesy of Professor Kellar Autumn).

are specific to the surfaces in question and even to the surface topographies. Usually adhesion is based on capillary forces as mediated by water between the surfaces [12]. However, for mobile devices the almost inherent surface tackiness should be avoided and it is desirable that the concepts do not depend on the details on the surfaces. Another bioinspired concept focuses on learning how small animals from insects up to kilogram-scale lizards can climb vertical walls. As the most spectacular example is related to the gecko [130–136], these adhesive concepts have been named after this lizard, see Figure 2.11. The underlying adhesion mechanism has been studied extensively. The dominant interaction involves van der Waals forces and requires that the adherent molecules are brought very close to each other in a consistent and adaptive way. To do this, the foot pad of the gecko contains about 14 000 hair-like setae, each with a diameter of ca. 5 µm. These are of submicrometer length and end in a set of flat spatulae. There-fore the foot surface has a hierarchical nanomicrostructure that is highly adaptive. In insects the hairs are thicker, and it is found that the heavier the animal that is to be suspended, the thinner are the hairs. Inspired by these observations, there have been many efforts to construct synthetic adhesive surfaces that imitate the gecko adhesion. Surface structures have been prepared by, e.g., lithographic methods [135] and using vertical multiwalled CNT [137] (see Figure 2.11(b), (c)). Such a surface measuring 4 mm × 4 mm has been demonstrated to carry a load of 1.5 kg.

2.5 Self-healing polymers

The final bioinspired concept concerns self-healing, which is ubiquitous in natural ma-terials to replace aged or worn parts of the biological organisms. In synthetic materials self-healing has been extensively pursued [9–11, 139–146], e.g., for reinforced compos-ites for the machine construction industry. There crack formation and propagation within the polymer matrix can be serious, and catastrophic crack growth has to be prevented. A classic approach has been to replace the fibrous reinforcements in composites with hollow fibrous tubings that contain a curing agent. Upon breaking, the curing agent is released. Even though the concept may be feasible in large machines, it would be rela-tively complicated to realize and easier methods would be welcome. A possible method is based on supramolecular polymers [10]. Polymers are chainlike, long molecules in which the repeat units are held together by strong chemical links (covalent bonds). Therefore, the polymer chains are in most cases stable, i.e. the intrachain linkages are permanent whereas interchain interactions depend on the polymer type and external conditions. A supramolecular polymer, in contrast, consists of repeat units that are con-nected together through nonpermanent physical interactions, such as hydrogen bonding, see Figure 2.12(a) [91, 147]. The physical bonds have to be strong enough to allow chainlike formation at room temperatures in order to allow the entangled structures that usually give the polymerlike properties. But the physical interactions between the repeat units can be designed to be weak enough to open reversibly on heating. When heated, the materials behave as simple fluids allowing easy processing, whereas on cooling down to the temperatures at which they are used the entangled chains reform and the material behaves like a solid polymer.

It has been observed that if one prepares branched repeat units with strongly physically interacting end-groups and where ordered packings are prohibited, reversible networks are formed that behave like rubbers [10]. A curious feature of these networks is that due to the high number of physical interaction sites and the soft plasticized material, the materials are self-healing: upon cutting the material and pressing the surfaces together again the parts reconnect, and the original materials properties are recovered within some hours. It is obvious that such materials could be a useful tool when trying to produce practical materials with self-healing properties and appropriate materials properties for use in future mobile devices.

(a) Supramolecular polymer

(b) Supramolecular self-healing elastomer

(c) Self-healing demonstrated

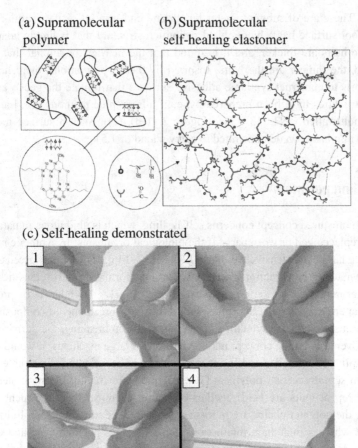

Figure 2.12 (a) Supramolecular polymer constructs confer new properties [91, 147]. The dotted lines depict dynamic hydrogen bonds between the hydrogen-bonding donor and acceptor sites which allows the formation of long chainlike entities. (b) Supramolecular elastomers allow self-healing [10]. In this case the hydrogen bonds lead to a networklike structure, mediated by hydrogen bonds. (c) Self-healing demonstrated. In part adapted with permission of Professor Ludvik Leibler; http://www2.cnrs.fr/en/1124.htm.

2.6   Conclusions

Our vision is that nanotechnology will provide an extensive toolkit enabling the tailoring of material at the molecular level to precise structures that lead to new well-defined functions. This bottom-up approach utilizing self-assembly of selected molecules and nanoscale building blocks is the basis for the development of novel specific materials. Several material concepts were reviewed to give some examples and to discuss the possibilities of nanotechnology and modern materials science for future mobile devices. These should be seen as possible paths towards the sought-for properties rather than as ready-made solutions. In practice, one of the key issues for many novel nanomaterials is controlling the material morphologies and properties at different scales, in particular combining the bottom-up building of materials with top-down fabrication methods in a practical way. The particular challenges involved in moving the concepts from the laboratories to manufactured products are related to economics, processing, reliability, reproducibility, lifetime, and systems compatibility.

References

[1] P. Ball, *Designing the Molecular World, Chemistry at the Frontier*, Princeton University Press, 1994.

[2] G. M. Whitesides and B. Grzybowski, Self-assembly at all scales, *Science*, **295**, 2418–2421, 2002.

[3] R. A. L. Jones, *Soft Machines: Nanotechnology and Life*, Oxford University Press, 2004.

[4] P. M. Ajayan *et al.*, eds., *Nanocomposite Science and Technology*, VCH-Wiley, 2004.

[5] G. A. Ozin and A. C. Arsenault, *Nanochemistry; A Chemical Approach for Nanomaterials*, The Royal Society of Chemistry, 2005.

[6] K. Matyjaszewski *et al.*, eds., *Macromolecular Engineering*, Wiley-VCH, 2007.

[7] P. Fratzl, Biomimetic materials research: what can we really learn from nature's structural materials?, *J. R. Soc. Interface*, **4**, 637–642, 2007.

[8] M. A. Meyers *et al.*, Biological materials: Structure and mechanical properties, *Prog. Mater. Science*, **53**, 1–206, 2008.

[9] S. D. Bergman and F. Wudl, Mendable polymers, *J. Mater. Chem.*, **18**, 41–62, 2008.

[10] P. Cordier *et al.*, Self-healing and thermoreversible rubber from supramolecular assembly, *Nature*, **451**, 977–980, 2008.

[11] R. P. Wool, Self-healing materials: a review, *Soft Matter*, **4**, 400–418, 2008.

[12] P.-G. de Gennes *et al.*, *Capillarity and Wetting Phenomena; Drops, Bubbles, Pearls, and Waves*, Springer-Verlag, 2002.

[13] R. Blossey, Self-cleaning surfaces – virtual realities, *Nat. Mat.*, **2**, 301–306, 2003.

[14] A. Tuteja *et al.*, Designing superoleophobic surfaces, *Science*, **318**, 1618–1622, 2007.

[15] D. Quéré, Wetting and roughness, *Annu. Rev. Mat. Res.*, **38**, 71–99, 2008.

[16] H. Gleiter, Our thoughts are ours, their ends none of our own: Are there ways to synthesize materials beyond the limitations of today?, *Acta Mat*, **56**, 5875–5893, 2008.

[17] M. C. Petty, *Molecular Electronics; from Principles to Applications*, Wiley, 2007.

[18] G. Hadziioannou and C. G. Malliaras, eds., *Semiconducting Polymers; Chemistry, Physics and Engineering*, Wiley-VCH, 2007.

[19] H. S. Nalwa, ed., *Handbook of Organic Electronics and Photonics*, American Scientific Publishers, 2008.

[20] N. Koch *et al.*, Special Issue: "Organic Materials for Electronic Applications", *Appl. Phys. A: Mater. Sci. Process*, **95**, 2009.

[21] Y. Lin *et al.*, Self-directed self-assembly of nanoparticle/copolymer mixtures, *Nature*, **434**, no. 3, 55–59, 2005.

[22] A. C. Balazs *et al.*, Nanoparticle polymer composites: where two small worlds meet, *Science*, **314**, 1107–1110, 2006.

[23] B. E. Chen *et al.*, A critical appraisal of polymer-clay nanocomposites, *Chem. Soc. Rev.*, **37**, 568–594, 2008.

[24] D. R. Paul and L. M. Robeson, Polymer nanotechnology: nanocomposites, *Polymer*, **49**, 3187–3204, 2008.

[25] G. M. Whitesides, Self-assembling materials, *Sci. Am.*, **273**, 146, 1995.

[26] F. S. Bates and G. H. Fredrickson, Block copolymers – designer soft materials, *Physics Today*, **52**, 32–38, 1999.

[27] I. W. Hamley, Nanotechnology with soft materials, *Angew. Chem. Int. Ed.*, **42**, 1692–1712, 2003.

[28] B. L. Zhou, Bio-inspired study of structural materials, *Mat. Sci. Eng. C*, **11**, 13–18, 2000.

[29] U. G. K. Wegst and M. F. Ashby, The mechanical efficiency of natural materials, *Philos. Mag.*, **84**, 2167–2186, 2004.

[30] M. F. Ashby, On the engineering properties of materials, *Acta Metall.*, **5**, 1273–1293, 1989.

[31] Z. Tang *et al.*, Nanostructured artifical nacre, *Nat. Mat.*, **2**, 413–418, 2003.

[32] P. Fratzl *et al.*, *J. Mat. Chem.*, **14**, 2115–2123, 2004.

[33] B. Fiedler *et al.*, Fundamental aspects of nano-reinforced composites, *Comp. Sci. Tech.*, **66**, 3115–3125, 2006.

[34] S. C. Tjong, Structural and mechanical properties of polymer nanocomposites, *Mater. Sci. Eng. R*, **53**, 73–197, 2006.

[35] P. Podsiadlo *et al.*, Ultrastrong and stiff layered polymer nanocomposites, *Science*, **318**, 80–83, 2007.

[36] D. W. Schaefer and R. S. Justice, How nano are nanocomposites?, *Macromolecules*, **40**, 8501–8517, 2007.

[37] M. Alexandre and P. Dubois, Nanocomposites, in *Macromolecular Engineering*, vol. 4, K. Matyjaszewski *et al.*, eds., pp. 2033–2070, Wiley-VCH, 2007.

[38] L. J. Bonderer *et al.*, Bioinspired design and assembly of platelet reinforced polymer films, *Science*, **319**, 1069–1073, 2008.

[39] L. A. Utracki, *Clay-Containing Polymer Nanocomposites*, Rapra Technology Ltd., 2004.

[40] J. Njuguna and K. Pielichowski, Polymer nanocomposites for aerospace applications: characterization, *Adv. Eng. Mater.*, **6**, 204–210, 2004.

[41] W. Wang *et al.*, Effective reinforcement in carbon nanotube-polymer composites, *Phil. Trans. R. Soc. A*, **366**, 1613–1626, 2008.

[42] A. Usuki *et al.*, Synthesis of nylon 6–clay hybrid, *J. Mater. Res.*, **8**, 1179–1184, 1993.

[43] K. Yano *et al.*, Synthesis and properties of polyimide-clay hybrid, *J. Polym. Sci., Part A: Polym. Chem.*, **31**, 2493–2498, 1993.

[44] A. Okada and A. Usuki, The chemistry of polymer-clay hybrids, *Mater. Sci. Eng. C*, **3**, 109–115, 1995.

[45] M. Okamoto *et al.*, A house of cards structure in polypropylene/clay nanocomposites under elongational flow, *Nano Letters*, **1**, no. 6, 295–298, 2001.

[46] S. S. Ray and M. Okamoto, Polymer/layered silicate nanocomposites: a review from preparation to processing, *Prog. Polym. Sci.*, **28**, 1539–1641, 2003.

[47] F. Gao, *Mater. Today*, **7**, 50, 2004.

[48] A. Usuki *et al.*, Polymer-clay nanocomposites, *Adv. Polym. Sci.*, **179**, 135–195, 2005.

[49] M. Okamoto, Polymer/layered filled nanocomposites: an overview from science to technology, in *Macromolecular Engineering*, vol. 4, K. Matyjaszewski, *et al.*, eds., pp. 2071–2134, Wiley-VCH, 2007.

[50] A. J. Patil and S. Mann, Self-assembly of bio-inorganic nanohybrids using organoclay building blocks, *J. Mat. Chem.*, **18**, 4605–4615, 2008.

[51] S. Srivastava and N. A. Kotov, Composite layer-by-layer (LBL) assembly with inorganic nanoparticles and nanowires, *Acc. Chem. Res.*, **41**, 1831–1841, 2008.

[52] P. Bordes *et al.*, Nano-biocomposites: biodegradable polyester/nanoclay systems, *Prog. Polym. Sci*, **34**, 125–155, 2009.

[53] O. L. Manevitch and G. C. Rutledge, Elastic properties of a single lamella of montmorillonite by molecular dynamics simulation, *J. Phys. Chem B*, **108**, 1428–1435, 2004.

[54] J.-H. Lee *et al.*, Properties of polyethylene-layered silicate nanocomposites prepared by melt intercalation with a PP-g-MA compatibilizer, *Composite Sci. Tech.*, **65**, 1996–2002, 2005.

[55] J. R. Capadona *et al.*, A versatile approach for the processing of polymer nanocomposites with self-assembled nanofiber templates, *Nat. Nanotech.*, **2**, 765–768, 2007.

[56] A. Dufresne, Polysaccharide nano crystal reinforced nanocomposites, *Can. J. Chem.*, **86**, 484–494, 2008.

[57] S. V. Ahir *et al.*, Polymers with aligned carbon nanotubes: Active composite materials, *Polymer*, **49**, 3841–3854, 2008.

[58] E. W. Wong *et al.*, Nanobeam mechanics: elasticity, strength, and toughness of nanorods and nanotubes, *Science*, **277**, 1971–1975, 1997.

[59] M.-F. Yu *et al.*, Carbon nanotubes under tensile load, *Science*, **287**, 637–640, 2000.

[60] M. Alexandre and P. Dubois, Polymer-layered silicate nanocomposites: preparation, properties and uses of a new class of materials, *Mat. Sci. Eng.*, **28**, 1–63, 2000.

[61] R. R. S. Schlittler *et al.*, Single crystals of single-walled carbon nanotubes formed by self-assembly, *Science*, **292**, 1139, 2001.

[62] A. B. Dalton *et al.*, Super-tough carbon-nanotube fibers, *Nature*, **423**, 703–2003.

[63] T. Fukushima *et al.*, Molecular ordering of organic molten salts triggered by single-walled carbon nanotubes, *Science*, **300**, 2072–2074, 2003.

[64] C. Wang *et al.*, Polymers containing fullerene or carbon nanotube structures, *Prog. Polym. Sci.*, **29**, 1079–1141, 2004.

[65] J. H. Rouse *et al.*, Polymer/single-walled carbon nanotube films assembled via donor–acceptor interactions and their use as scaffolds for silica deposition, *Chem. Mater.*, **16**, 3904–3910, 2004.

[66] T. Liu *et al.*, Morphology and mechanical properties of multiwalled carbon nanotubes reinforced nylon-6 composites, *Macromolecules*, **37**, 7214–7222, 2004.

[67] M. Zhang *et al.*, Strong, transparent, multifunctional, carbon nanotube sheets, **309**, 1215–1219, 2005.

[68] L. Jiang and L. Gao, Fabrication and characterization of ZnO-coated multi-walled carbon nanotubes with enhanced photocatalytic activity, *Mat. Chem. Phys.*, **91**, 313–316, 2005.

[69] J. N. Coleman *et al.*, Mechanical reinforcement of polymers using carbon nanotubes, *Adv. Mat.*, **18**, 689–706, 2006.

[70] M. Olek *et al.*, Quantum dot modified multiwall carbon nanotubes, *J. Phys. Chem. B*, **110**, 12901–12904, 2006.

[71] M. Moniruzzaman *et al.*, Tuning the mechanical properties of SWNT/nylon 6,10 composites with flexible spacers at the interface, *Nano Lett.*, **7**, 1178–1185, 2007.

[72] L. Bokobza, Multiwall carbon nanotube elastomeric composites: a review, *Polymer*, **48**, 4907–4920, 2007.

[73] Y. Liu *et al.*, Noncovalent functionalization of carbon nanotubes with sodium lignosulfonate and subsequent quantum dot decoration, *J. Phys. Chem. C*, **111**, 1223–1229, 2007.

[74] Y. Ye *et al.*, High impact strength epoxy nanocomposites with natural nanotubes, *Polymer*, **48**, 6426–6433, 2007.

[75] P. Miaudet *et al.*, Hot-drawing of single and multiwall carbon nanotube fibers for high toughness and alignment, *Nano Lett.*, **5**, 2212–2215, 2005.

[76] R. Haggenmueller *et al.*, Interfacial in situ polymerization of single wall carbon nanotube/ nylon 6,6 nanocomposites, *Polymer*, **47**, 2381–2388, 2006.

[77] K. Fleming *et al.*, Cellulose crystallites, *Chem. Eur. J.*, **7**, 1831–1835, 2001.

[78] D. Klemm *et al.*, Cellulose: fascinating biopolymer and sustainable raw material, *Angew. Chem. Int. Ed.*, **44**, 3358–3393, 2005.

[79] M. Nogi *et al.*, Optically transparent nanofiber paper, *Adv. Mat.*, **20**, 1–4, 2009.

[80] J. R. Capadona *et al.*, Stimuli-responsive polymer nanocomposites inspired by the sea cucumber dermis, *Science*, **319**, 1370–1374, 2008.

[81] M. Henriksson *et al.*, An environmentally friendly method for enzyme-assisted preparation of microfibrillated cellulose (MFC) nanofibers, *Eur. Polym. J.*, **43**, no. 8, 3434–3441, 2007.

[82] M. Pääkkö *et al.*, Enzymatic hydrolysis combined with mechanical shearing and high-pressure homogenization for nanoscale cellulose fibrils and strong gels, *Biomacromolecules*, **8**, 1934–1941, 2007.

[83] M. Pääkkö *et al.*, Long and entangled native cellulose I nanofibers allow flexible aerogels and hierarchically porous templates for functionalities, *Soft Matter*, **4**, 2492–2499, 2008.

[84] M. Henriksson *et al.*, Cellulose nanopaper structures of high toughness, *Biomacromolecules*, **9**, 1579–1585, 2008.

[85] A. N. Nakagaito and H. Yano, Novel high-strength biocomposites based on microfibrillated cellulose having nano-order-unit web-like network structure, *Appl. Phys. A*, **80**, 155–159, 2005.

[86] M. Nogi and H. Yano, Transparent nanocomposites based on cellulose produced by bacteria offer potential innovation in the electronics device industry, *Adv. Mat.*, **20**, 1849–1852, 2008.

[87] T. Zimmerman *et al.*, Cellulose fibrils for polymer reinforcement, *Adv. Eng. Mater.*, **6**, no. 9, 754–761, 2004.

[88] F. J. M. Hoeben *et al.*, About supramolecular assemblies of π-conjugated systems, *Chem. Rev.*, **105**, 1491–1546, 2005.

[89] J. Ruokolainen *et al.*, Switching supramolecular polymeric materials with multiple length scales, *Science*, **280**, 557–560, 1998.

[90] S. Valkama *et al.*, Self-assembled polymeric solid films with temperature-induced large and reversible photonic bandgap switching, *Nat. Mat.*, **3**, 872–876, 2004.

[91] R. P. Sijbesma *et al.*, Reversible polymers formed from self-complementary monomers using quadruple hydrogen bonding, *Science*, **278**, 1601–1604, 1997.

[92] T. Thurn-Albrecht *et al.*, Ultrahigh-density nanowire arrays grown in self-assembled copolymer templates, *Science*, **290**, 2126–2129, 2000.

[93] C. Tang *et al.*, Evolution of block copolymer lithography to highly ordered square arrays, *Science*, **322**, 429–432, 2008.

[94] S. Park *et al.*, Macroscopic 10-terabit-per-square-inch arrays from block copolymers with lateral order, *Science*, **323**, 1030–1033, 2009.

[95] M. Muthukumar *et al.*, Competing interactions and levels of ordering in self-organizing polymeric materials, *Science*, **277**, 1225–1232, 1997.

[96] A. Pron and P. Rannou, Processible conjugated polymers: from organic semiconductors to organic metals and superconductors, *Prog. Polym. Sci.*, **27**, 135–190, 2002.

[97] P. Leclere *et al.*, Supramolecular organization in block copolymers containing a conjugated segment: a joint AFM/molecular modeling study, *Prog. Polym. Sci.*, **28**, 55–81, 2003.

[98] V. Percec *et al.*, Controlling polymer shape through the self-assembly of dendritic side-groups, *Nature*, **391**, 161–164, 1998.

[99] X. Feng *et al.*, Towards high charge-carrier mobilities by rational design of the shape and periphery of discotics, *Nat. Mat.*, **8**, 421–426, 2009.

[100] T. Kato *et al.*, Functional liquid-crystalline assemblies: self-organized soft materials, *Angew. Chem. Int. Ed. Engl.*, **45**, 38–68, 2006.

[101] G. Holden *et al.*, eds., *Thermoplastic Elastomers*, Hanser, 2004.

[102] B. O'Regan and M. Graetzel, A low-cost, high-efficiency solar cell based on dye-sensitized colloidal TiO_2 films, *Nature*, **353**, 737–740, 1991.

[103] M. Gratzel, Dye-sensitized solid-state heterojunction solar cells, *MRS Bulletin*, **30**, 23–27, 2005.

[104] E. J. W. Crossland *et al.*, A bicontinuous double gyroid hybrid solar cell, *Nano Lett.*, in press, 2009.

[105] D. L. Kaplan *et al.*, Self-organization (assembly) in biosynthesis of silk fibers – a hierarchical problem, *Mat. Res. Soc. Symp. Proc.*, **255**, 19–29, 1992.

[106] D. L. Kaplan *et al.*, Silk, in *Protein-Based Materials*, K. McGrath, D. Kaplan, eds., pp. 103–131, Birkhauser, 1997.

[107] J. M. Gosline *et al.*, The mechanical properties of spider silks: from fibroin sequence to mechanical function, *J. Exp. Biol.*, **202**, 3295–3303, 1999.

[108] S. Kubik, High-performance fibers from spider silk, *Angew. Chem. Int. Ed.*, **41**, 2721–2723, 2002.

[109] R. Valluzzi *et al.*, Silk: molecular organization and control of assembly, *Phil. Trans. R. Soc. London, Ser. B: Bio. Sci.*, **357**, 165–167, 2002.

[110] F. Vollrath and D. Porter, Spider silk as archetypal protein elastomer, *Soft Matter*, **2**, 377–385, 2006.

[111] J. G. Hardy *et al.*, Polymeric materials based on silk proteins, *Polymer*, **49**, 4309–4327, 2008.

[112] S. Mann *et al.*, *Biomineralization, Chemical and Biochemical Perspectives*, VCH Publishers, 1989.

[113] M. Darder *et al.*, Design and preparation of bionanocomposites based on layered solids with functional and structural properties, *Mat. Sci. Tech.*, **24**, 1100–1110, 2008.

[114] H.-O. Fabritius *et al.*, Influence of structural principles on the mechanics of a biological fiber-based composite material with hierarchical organization: the exoskeleton of the lobster *Homarus americanus*, *Adv. Mat.*, **21**, 391–400, 2009.

[115] G. Decher, Fuzzy nanoassemblies: towards layered polymeric multicomposites, *Science*, **277**, 1232–1237, 1997.

[116] N. C. Seeman, Nucleic acid junctions and lattices, *J. Theor. Biol.*, **99**, 237–247, 1982.

[117] E. Winfree *et al.*, Design and self-assembly of two-dimensional DNA crystals, *Nature*, **394**, 539–544, 1998.

[118] N. C. Seeman, DNA in a material world, *Nature*, **421**, 427–431, 2003.

[119] P. W. K. Rothemund, Folding DNA to create nanoscale shapes and patterns, *Nature*, **440**, 297–302, 2006.

[120] H. Yan *et al.*, DNA-templated self-assembly of protein arrays and highly conductive nanowires, *Science*, **301**, 1882–1884, 2003.

[121] E. S. Andersen *et al.*, DNA origami design of dolphin-shaped structures with flexible tails, *ACS Nano*, **2**, 1213–1218, 2008.

[122] A. Kuzyk *et al.*, Dielectrophoretic trapping of DNA origami, *Small*, **4**, 447–450, 2008.

[123] A. Lafuma and D. Quere, Superhydrophobic states, *Nat. Mat.*, **2**, 457–460, 2003.

[124] M. Callies and D. Quere, On water repellency, *Soft Matter*, **1**, 55–61, 2005.

[125] X. Feng and L. Jiang, Design and creation of superwetting/antiwetting surfaces, *Adv. Mat.*, **18**, 3063–3078, 2006.

[126] M. Ma *et al.*, Decorated electrospun fibers exhibiting superhydrophobicity, *Adv. Mat.*, **19**, 255–259, 2007.

[127] L. Gao *et al.*, Superhydrophobicity and contact-line issues, *MRS Bulletin*, **33**, 747–751, 2008.

[128] R. Fürstner *et al.*, Der Lotus-Effekt: Selbstreinigung mikrostrukturierter Oberflächen, *Nachricten aus der Chemie*, **48**, 24–28, 2000.

[129] P. Wagner *et al.*, Quantitative assessment to the structural basis of water repellency in natural and technical surfaces, *J. Exp. Bot.*, **54**, 1295–1303, 2003.

[130] K. Autumn *et al.*, Adhesive force of a single gecko foot-hair, *Nature*, **405**, 681–685, 2000.

[131] A. K. Geim *et al.*, Microfabricated adhesive mimicking gecko foot-hair, *Nat. Mat.*, **2**, 461–463, 2003.

[132] B. Bhushan, Adhesion of multi-level hierarchical attachment systems in gecko feet, *J. Adh. Sci. Tech.*, **21**, 1213–1258, 2007.

[133] B. N. J. Persson, Biological adhesion for locomotion on rough surfaces: basic principles and a theorist's view, *MRS Bulletin*, **32**, 486–490, 2007.

[134] K. Autumn and N. Gravish, Gecko adhesion: evolutionary nanotechnology, *Phil. Trans. R. Soc. A*, **366**, 1575–1590, 2008.

[135] A. del Campo and E. Arzt, Fabrication approaches for generating complex micro- and nanopatterns on polymeric surfaces, *Chem. Rev.*, **108**, 911–945, 2008.

[136] M. T. Northen *et al.*, A gecko-inspired reversible adhesive, *Adv. Mat.*, **20**, 3905–3909, 2008.

[137] L. Qu *et al.*, Carbon nanotube arrays with strong shear binding-on and easy normal lifting-off, *Science*, **322**, 238–242, 2007.

[138] J. Genzer and K. Efimenko, Recent developments in superhydrophobic surfaces and their relevance to marine fouling: a review, *Biofouling*, **22**, 339–360, 2006.

[139] S. Weiner *et al.*, Materials design in biology, *Mat. Sci. Eng. C*, **11**, 1–8, 2000.

[140] X. Chen *et al.*, A thermally re-mendable cross-linked polymeric material, *Science*, **295**, 1698–1702, 2002.

[141] M. Chipara and K. Wooley, Molecular self-healing processes in polymers, *Mat. Res. Soc. Symp. Proc.*, **851**, 127–132, 2005.

[142] A. C. Balazs, Modeling self-healing materials, *Materials Today*, **10**, 18–23, 2007.

[143] M. W. Keller *et al.*, Recent advances in self-healing materials systems, in *Adaptive Structures*, D. Wagg, ed., pp. 247–260, J. Wiley & Sons Ltd, 2007.

[144] R. S. Trask *et al.*, Self-healing polymer composites: mimicking nature to enhance performance, *Bioinspiration & Biomimetics*, **2**, P1–P9, 2007.

[145] D. Montarnal *et al.*, Synthesis of self-healing supramolecular rubbers from fatty acid derivatives, diethylene triamine, and urea, *J. Polym. Sci., Part A*, **46**, 7925–7936, 2008.

[146] S. Burattini *et al.*, A novel self-healing supramolecular polymer system, *Faraday Discuss. Chem. Soc.*, in press, 2009.

[147] R. P. Sijbesma and E. W. Meijer, Quadruple hydrogen bonded systems, *Chem. Commun.*, 5–16, 2003.

3 Energy and power

M. Rouvala, G. A. J. Amaratunga, D. Wei, and A. Colli

3.1 Energy needs of future mobile devices

3.1.1 Renewable energy – portable energy sources

The increasing need for portable energy storage density due to the growing number of miniaturized and thin devices is driving the current development of energy storage in phones and other portable devices [1]. Figure 3.1 shows the development from the mobile phones of the 1990s to the multimedia centers of two decades later, and the evident development of the devices to thinner and more flexible forms.

The total power consumption will become even more important when more electronic devices are embedded in the environment. Also with standalone devices designed to operate without mains power supply for long periods, like years, there are new requirements. Energy storage and power management are among the top three issues for customers and developers in current and future mobile multimedia portable devices.

Improvements in conventional battery at the current yearly level are not expected to provide enough energy density to meet all the requirements of future multimedia portables. Even though the use of cellular radio frequency (RF) engine power is expected to reduce with integrated circuit (IC) process and intelligent circuit development, the increasing number of radios and integration of new digital radio, video, and multimedia broadcasting (DAB, DVB-H, DMB) and channel decoders represents a significant challenge both in terms of energy consumption and component costs. Adding wireless local area network (WLAN) and local radio capabilities increases the overall power consumption of smartphones that are already suffering from high energy drain due to the high power consumption of 2.5G/3G wireless modules. Lower power dissipation will be needed in all technologies, through wafer fabrication, software, displays, embedded processors, and interconnects. In the longer term lower energy usage patterns will have some effect.

Energy density and power density mark two different perspectives in mobile energy. Energy is needed for long phone calls, ever increasing internet uploads with mobile radio technologies, and generally for the long stand-by times required by users. High-power sources on the other hand are needed to source fast bursts of current used by multimedia

Nanotechnologies for Future Mobile Devices, eds. T. Ryhänen, M. A. Uusitalo, O. Ikkala, and A. Kärkkäinen. Published by Cambridge University Press. © Cambridge University Press 2010.

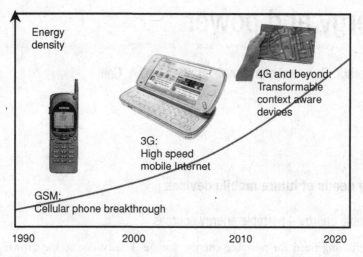

Figure 3.1 Roadmap of the required energy density in mobile devices.

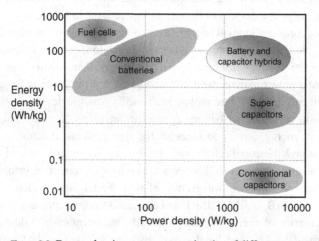

Figure 3.2 Energy density versus power density of different power sources for mobile devices.

applications on devices: cameras with flashes, hard disk drives (HDDs), high-resolution displays etc.

The comparison of different energy storage technologies is presented in Figure 3.2.

Mobile device battery technologies have developed towards reducing size and weight, and having longer and more stable performance. Increased energy density is needed to implement energy storage in smaller packages enabling new form factor devices, like very thin phones or accessories. Typically, the consumer markets require low cost, which applies especially to very-high-volume products and emerging markets.

Typical achievable numbers with current technologies are listed below:

- weight: <50 g;
- size: <50 × 30 × 5 mm [l × w × h];
- energy density: 250–500 Wh/kg, and towards 600 Wh/kg in 2010;
- low cost $5/1000 mA h.

In addition reduced charging time is a complementary goal for energy density.

Supercapacitors are very desirable for power storage but, as they have low energy densities compared to the batteries and fuel cells, they are not a favorable solution to the energy problem. Lithium ion cells are lighter and have higher energy density, power density, and voltage than their old competitors (NiCd, NiMH and lead acid). As lithium is known to be one of the most electronegative materials it is difficult to find competitive alternatives for use in electrochemical cells. High electronegativity has the advantage of applying fewer cells in series to create the needed operating voltage of the device.

3.1.2 Power consumption in a mobile phone – now and in the future

Most of the energy used by a mobile phone is consumed when it is on standby. Typical standby power consumption is 18–20 mW. At the other end of power usage is a 3G handset enabled video conferencing, which uses almost all the functionalities of a smart phone – camera, display, and radio interface, with possibly also an integrated hands-free functionality, i.e. a loudspeaker. The average power is then increased to 2000 mW; peaks over 5 W are possible when high-speed data interfaces (e.g., enhanced data rates for GSM evolution (EDGE)) are initiated. A yearly battery energy density increase of 8% should not be seen as increased battery capacity as such, because of the simultaneous reduction in size. Battery capacity has stabilized at around 1000 mA h for high-end products, despite the energy density increase. Battery thickness has fallen from 6–8 mm in 2003 to 3–5 mm in products developed after 2006 (see Figure 3.3). This has mostly been due to pressure from device thinness. In 2006 Samsung produced 12 mm thick products, and since then NEC, Nokia and other technology leaders have been able to develop phones with a maximum thickness of less than 10 mm. The average thickness of phones in 2007 was 15.8 mm.

There are three major energy drains in a modern mobile phone: displays, processing power (semiconductors in general), and the RF front-end (including the aerial interface). Each of these uses roughly one third of the total power.

Large displays (2–3 inches diagonally) use 211–272 mW. The trend is for increased display power. Even though the high-speed interfaces of high-resolution displays are being developed towards serialized low-power interconnects, the display resolution and number of pixels are still growing. Using a phone in direct sunlight still raises the need for improved brightness. The brightness in thin-film transistor (TFT) liquid crytal displays (LCDs) is generated by white LEDs, typically consuming 6–12 mA of current. In general the LED current consumes the majority of display power.

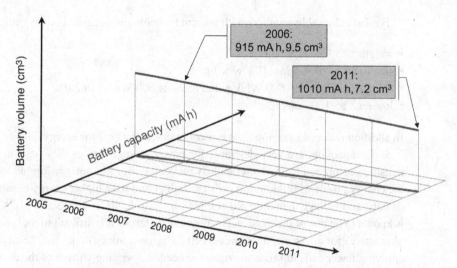

Figure 3.3 Trend of the capacity and size of batteries used for mobile phones.

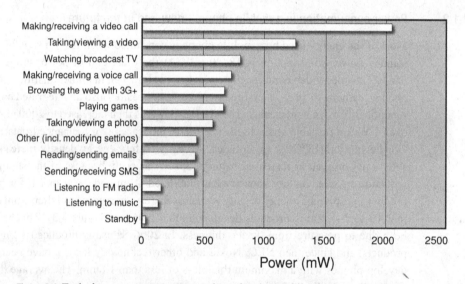

Figure 3.4 Typical average power consumption related to mobile phone applications.

The RF front-end consists of an antenna, a power amplifier, and a receiver. The average receiver power use has generally been small, between 50 and 70 mW for 2G and 3G systems. In the future (3G LTE, WiMax) the implementation of the receiver will have to meet even more challenging requirements for the energy sources, storage, and management, due to wider bandwidth radio interfaces, increased complexity in coding, use of fast Fourier transforms, and a greater number of supported radio standards. Radio interface power use is a function of distance from the base station, power amplifier (PA) efficiency, and data rate.

The average power consumption calculations in Figure 3.4 are based on analysis of the required functionalities, time, and the percentage of full power needed from each component. Components include: display and backlight, camera, RF and modem, audio, application processor and memory.

As is evident in Figure 3.4 most power-hungry applications are those related to multimedia, video calls, receiving TV broadcasts, and web browsing. These are also the needs for the future devices with increasing need for power bursts, in addition to long-term standby power need.

3.1.3   Nanotechnologies for energy

There are many ways in which nanotechnologies can play a role in future energy solutions. The fact that nanotechnology can be used to create very strong and light materials will have the end benefit of reducing the energy consumed in transport. More directly, a variety of materials in nanoparticle form can be used to develop enhanced energy storage and harvesting solutions, making viable the generation and storage of energy locally on a mobile platform. Hydrogen generation from water with sunlight or photocatalytic methanol production for fuel cells could, e.g., address the problem of having to carry around a larger battery to access a full feature set on the phone.

Nanotechnology holds the promise of allowing very-low-cost fuel cells to transform the fuel into electrical energy. As noted above the LED back light in the mobile phone display consumes a significant portion of the overall power available. New generation nanoparticle-based LEDs, coupled with nanophotonic features to direct the light will increase the power use efficiency of the mobile phone display. Nanostructured photovoltaic cells which can absorb a wide spectrum of energies, cathodes and anodes which exhibit enhanced electrochemical and catalytic activity for batteries and fuel cells are some of the other exciting possibilites.

3.2   Basics of battery and power source technologies

3.2.1   Batteries and supercapacitors

Batteries and capacitors differ fundamentally in the way that the charge transfer takes place in the charging/discharging processes. Batteries charge with Faradaic charge transfer processes, in which electron transfer takes place at the electrode surface. In a battery every Faradaic charge carrier is neutralized at the electrode surface by charge transfer. In common physical capacitors, the charge storage process is non-Faradaic and electrostatic in nature, i.e., ideally no electron transfer takes place at the electrodes. In such capacitors, the charge storage is accumulated between an insulating dielectric (molecular layer, a vacuum, mica separator, etc.) in the anode and cathode electrostatically. In batteries the ion transfer/electron transfer results in chemical changes in electrode materials due to the consequent change of oxidation state, hence leading to a change in the chemistry of electroactive materials.

The growing need for portable energy storage density brought about by miniaturization and the slim form factor requirement is driving the current development of lithium batteries as well as supercapacitors. In a true supercapacitor, which is also referred to as an electrochemical capacitor, the charge is accumulated in electrical double layers at the electrode–electrolyte interface.

There are two types of supercapacitors, depending on the charge storage mechanism [2]. In the electrochemical double layer capacitor (EDLC), a pure electrostatic attraction occurs between the ions and the charged surface of the electrode which is generally activated carbon. The electrons involved in double layer charging are the delocalized conduction band electrons of the carbon electrode. In the second type of capacitor, pseudocapacitors, electrons are additionally involved in quick Faradaic reactions and are transferred to or from the valence bands of the redox cathode or anode reagent, although they may arrive in or depart from the conduction band of carbon. Electrode materials with pseudocapacitance properties are generally metal oxides (RuO_2, MnO_2 etc.) or conducting polymers.

The EDLC consists of two electrodes, with electrolyte–dielectric between them. The electric double layer at each electrode includes a compactly packed layer and a surrounding wider layer. The compactly packed layer is about 0.5–0.6 nm thick, depending on the size of the ion charge carriers, and the 1–100 nm thick wider layer consists of solvent molecules and ions and contributes to the charge storage. Because of this ultimate thinness very high *capacitances can be achieved*. Supercapacitors based on EDLC capacitance differ from normal capacitors in the action of the dielectric medium. In a normal capacitor the bulk phase of the electrolyte dielectric medium defines the capacitance, but in an EDLC the microscopic properties of the dielectric determine the specific capacitance.

3.2.2   Nanostructured electrodes

Because of the elemental properties of lithium i.e. lightness and high electronegativity, it is used as the de facto standard material in almost all cellular phones and digital cameras today. For a lithium battery, the standard material for the negative electrode is graphite. Its typical structure is shown in Figure 3.5. In the lithium ion battery charging process, charge carriers extracted from cathode metal oxide diffuse through the electrolyte in a Faradaic process, and are adsorbed or intercalated on or near the surface, as shown in Figure 3.5. In the discharging process the flow direction is reversed and the current is applied to the load.

Graphite is naturally porous and has been well studied for capturing lithium through adsorption and intercalation. Since the synthesis of fullerenes, nanotubes, and other carbon nanostructures, carbon nano-onions (CNOs) and carbon nanohorns (CNHs), these materials have been explored for their high adsorption capabilities for hydrogen storage, and as active materials as membranes for fuel cells and batteries [3]. Carbon nanotube (CNT) EDLCs are already in commercial production. CNT batteries do not exist as yet. Some reasons for this are listed below.

The disadvantage of nanostructured electrodes is that undesired electrode–electrolyte reactions can occur leading to:

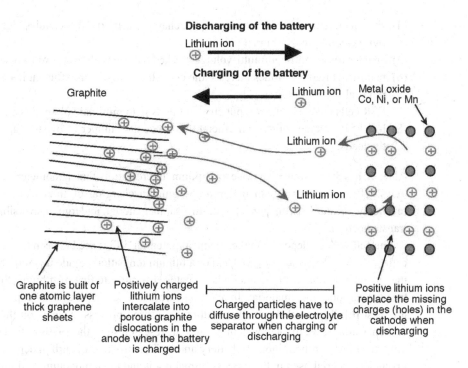

Figure 3.5 Charging and discharging processes of a nanostructured lithium ion battery: (a) the porous graphite anode, and (b) the metal oxide cathode. The lithium ions diffuse through the electrolyte during the charging and discharging.

(1) increased self-discharge to electrolyte due to higher surface area of the nanostructures;
(2) poor cycle life and calendar life due to unwanted electrode reactions of large surface areas;
(3) safety concerns that arise from primary nanoparticle packing (dendrites), leading to possible local heating and short circuits between the electrodes;
(4) agglomeration of nanoparticles and thus lower volumetric energy density;
(5) secondary nanoparticle formation (unwanted nanoparticle aggregates) that can lead to a lower surface area than wanted, and thus lead to capacity and durability losses.

In addition, the potentially more complex synthesis makes industrial production of nanomaterials difficult.

However, there are many advantages of nanostructured electrodes over conventional structures [4]:

(1) better accommodation of the strain associated with insertion and removal of charge carrier ions (lithium ions in lithium batteries) – this gives improved cycle life;
(2) new reactions are possible in comparison with bulk materials;
(3) greater contact/surface area (SSA) for the electrode–electrolyte surface – this enables high EDLC and low equivalent series resistance (ESR);

(4) short path lengths for electron transport/charge transport – this enables higher resistivity electrolytes, or higher power;

(5) enabler for a multi-cell/multi-voltage stacked battery with less power noise;

(6) nanostructured electrodes increase the cycle life due to more efficient insertion and removal of the lithium ions;

(7) ESR cells enable higher output current bursts with minimal voltage loss;

(8) low ESR voltage source can efficiently reduce the number of bypass capacitors at the loads.

Also, it has been shown in case of lithium intercalation that nanometer scale particles (20 nm haematite) can undergo charging–discharging or intercalation cycling reversibly, while 1–2 μm particles of the same material undergo irreversible phase transformation.

The cell voltage depends on the chemical potential difference between the anode and cathode $V = -(\mu_{cathode} - \mu_{anode})/e$. For a lithium ion battery anode it is important that the chemical potential of the anode host material is close to that of lithium [5]. This is true for carbonaceous material. In addition carbons have a high cycling capability and low cost [6]. The mechanism of the lithium insertion process depends on the type of carbon. Carbon has, in fact, been under serious research since the 1970s to find the right kind of carbon for the lithion ion battery anode. The structure of carbon depends on the precursor material used in the process of making it and the temperature and pressure at which it is processed.

Currently used carbons can be classified in three classes: (1) graphitic carbons, prepared by heating soft carbon precursors above 2400 °C (2) soft and hard carbons; heated to 500–700 °C, with substantial hydrogen content; and (3) hard carbons made up of single graphene layers or spheres. The voltage profiles of the three different classes are remarkably different, suggesting different reaction mechanisms in the process of charging and discharging [7]. Class 1 graphitic carbons go through a phase change of the turbostratic graphitic layers of the material, which leads to small dimensional changes, rotation and disorder in graphene layers of the graphite. Class 3 hard carbons have very little hydrogen content. In this class the lithium insertion is believed to happen reversibly between the graphene layers and onto internal surfaces of the nanopores and micropores.

Most common graphite anodes in commercial products today are made of class 1 graphites. Silicon has been proposed as the anode material, but this is still in its infancy compared to graphite. Silicon has a large alloying capacity for lithium [8–10].

3.2.3   Carbon nanohorns (CNHs) and their special surface properties

CNHs consist of two-dimensional graphene sheets and have special physical properties because of their apical disinclination in nanocones, and specific energy dispersion related to curve surfaces [11–13].

CNHs shown in Figure 3.6, have been studied for their chemical properties regarding adsorption, catalysis, and functionalization, but individual nanocones have not been

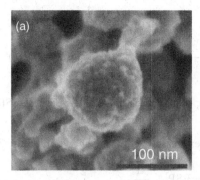
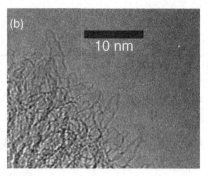

Figure 3.6 Two different kinds of carbon nanohorn (CNH) structures formed by two-dimensional graphene sheets: (a) bud-like carbon nanohorns; (b) dahlia-like carbon nanohorns. (Nokia Research Center and University of Cambridge).

examined in detail. Nanocones can have the special property of carrying molecules and nanoparticles which could be a beneficial property in nanobiology and medicine. The main applications for CNH aggregates seen today include fuel cell electrodes and supercapacitor–battery structures for high specific surface area (SSA) and surface corrugation. These surface properties are especially applicable to electrodes of EDLCs [14].

Charging the EDLC electrode involves charge density and distribution changes of the delocalized electrons, approximately one per atom. The electron wave functions at the conductive carbon–metal electrode decrease exponentially from the surface to the electrolyte, creating significant negative polarization at the surface, an overspill effect. These wave functions can be enhanced by negative polarization of the electrode. When polarizing the electrode positively the electron wave functions experience a higher boundary and electron density retreats towards the carbon–metal atomic centres at the boundary. The electron density is pushed in towards the electrode material compared to the neutral or negatively polarized cases. The outwardly pushed electron density areas interact more easily with the ions and charge carriers in the electrolyte in the compact layer (inner Helmholtz and outer Helmholtz layers). The variation in the surface electron density overspill affects the adsorption processes and other surface interaction and organization processes. Polarization of the surface in this way is similar in nature to electronic work function ϕ variations at the surface.

3.2.4 Supercapacitors – effect of surface

Equivalent series resistance (ESR) in batteries or capacitors is due to the series resistive elements in the cell itself and in the cell packaging. When resistance is built in electrolyte or electrode lead contact resistances, the resistance is, in fact, a real resistance element, instead of an equivalent. The formal equivalent series resistance is a resistance originating from kinetically limited relaxation processes in the dielectric itself. The formation of ESR in EDLC is due to distributed dielectric and interparticle resistances in the porous electrode matrix. The ideal impedance of a capacitor can be expressed as $Z(\omega) = j\omega C$

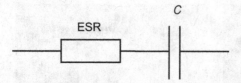

Figure 3.7 Equivalent circuit of a nonideal capacitor with a series resistive element, ESR.

and with the series resistive element as expressed below and shown in Figure 3.7:

$$Z(\omega) = \text{ESR} + \frac{1}{j\omega C}. \tag{3.1}$$

In capacitors ESR gives rise to a phase angle, which is less than 90° for the real and imaginary parts of the impedance. ESR is also often discussed as a series resistance element independent of its real or equivalent origin.

Minimization of ESR is a very fundamental need for supercapacitors as well as for batteries. In a capacitor ESR limits the power performance – the highest available power for charging and discharging is limited by the ESR.

The basic equation for capacitance is expressed using the dielectric constant ε between the electrodes, the electrode parallel surface area A, and the distance between the electrodes d:

$$C = \frac{\varepsilon \cdot A}{d}. \tag{3.2}$$

Supercapacitors utilize higher-surface-area electrodes and thinner dielectrics to achieve greater capacitances. This gives energy densities greater than those of conventional capacitors and power densities greater than those of batteries. With higher power density compared to batteries, supercapacitors may become an attractive power solution for mobile energy storage and delivery. They could also be used for distributed storage of harvested energy and local power delivery.

The fastest available power from a capacitor can be expressed with a RC time constant:

$$t_{\text{cutoff}} = R_{\text{ESR}} \cdot C. \tag{3.3}$$

When applying power and voltage equations $P = UI$, $U = RI$ and the capacitor energy equation to (3.3), we can express the power to energy relation in terms of the time constant of the capacitor:

$$\frac{P}{E} = \frac{2}{t_{\text{cutoff}}}. \tag{3.4}$$

This indicates that the power available from a capacitor depends on the energy in inverse relation to the RC time constant.

Also, importantly, this tells us that P/E ratio is a function of the physical construction and materials of the capacitor, while the actual capacitance and voltage range do not enter this equation as variables.

For maximum power $R_L = R_{ESR}$ thus becomes:

$$P_{max} = \frac{V_0^2}{4 \cdot R_{series}}.$$ (3.5)

In addition, it has to be noted that real EDLC capacitors have two electrodes, as do batteries. The series connection of capacitors seen at the cell electrodes is an inverse sum of the capacitances of the two EDLC layers (at the battery or the capacitor electrodes):

$$\frac{1}{C} = \frac{1}{C_{anode}} + \frac{1}{C_{cathode}}.$$ (3.6)

According to this statement, the electrode with smaller capacitance will contribute more to the total capacitance of the circuit.

For carbon electrode capacitors ESR has been shown to originate from a resistive layer at the surface of the carbon material [15, 16]. It is due to the blocking of the pores of the porous carbon electrode. Other than that, generally, the key factors limiting the power density and frequency response of a supercapacitor are the internal resistivity of an electrode, the contact resistivity between the electrode and the current collector, and the resistivity of the electrolyte in the pores of the electrode. In a graphite electrode this mostly means micropores. Reduction of these elements has been attempted. Entangled carbon nanotube mats have been proposed as electrodes, as well as using polished nickel foil to lower the contact resistance with the CNTs. CNT heat treatment at high temperature has been proposed to reduce the internal resistance in the CNT network created at the surface of an electrode. CNTs have also been grown directly onto metal current collectors. This was achieved with hot filament plasma enhanced chemical vapor deposition (HFPECVD). Highly concentrated colloidal suspensions of CNTs have been shown to reduce the ESR, as does a method of preparing multiwall carbon nanotube (MWNT) film with electrophoretic deposition (EPD). All in all, none of the methods proposed thus far has resulted in a high-volume method of production for CNT/MWNT networks, or other carbon nanostructure networks.

In batteries, ESR at the electrodes causes heat generation and voltage drop. The operational time of a device depends on the battery voltage drop. The flatter the voltage drop is, the longer the current can be sourced from the battery, before the voltage at the load drops under the V_{cutoff}, at which load circuits cease to function, because the sourced voltage V_{cc} is below the circuit functional limit V_{ccmin}. For short power bursts the voltage drop ΔV at the battery is dependent on the ESR as shown in Figure 3.8. This is a simplified schematic of constant current drain to the load. In reality, current bursts are also sourced from the battery at later stages of discharging. When the voltage in the load circuit drops close to V_{cutoff}, or V_{ccmin}, all the other current spikes start to make a difference, and when the sourced voltage is close to the V_{cutoff}, small current spikes can trip the voltage below the V_{cutoff} level, even though the long-term current would stay above it. With low ESR current sources these voltage ripples caused by sourced current are smaller. Hence, another advantage of a low ESR energy source is a longer battery lifetime due to smaller V_{cc} noise peaks during the later stages of the battery lifetime.

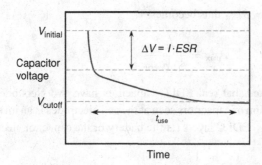

Figure 3.8 Discharging of a capacitor with an ESR (see Figure 3.7).

3.2.5   Porous electrodes

Chimiola *et al.* [17] found an anomalous capacitance increase for a carbon surface. The remarkable finding was that pores that are slightly smaller than solvated ions were the most effective in capacitance increase. The result was explained by distortion of solvation shells in small pores. This would allow the ion center to approach closer to the carbon electrode surface, thus giving the higher and improved capacitive effect.

Porosity analysis is needed to determine the adsorption process isotherms and the speed of the processes. The nature of the porosity defines factors such as: pore-size distribution or potential energy distribution, average pore size, shapes of the pores, relative positions of the pores, surface chemistry of the pores and diffusion paths controlling the rates of the adsorption.

3.2.6   Fuel cell technologies

Fuel cells for portable devices are approaching commercialization. Micro fuel cells have indeed developed significantly since the late 1990s, but still have technological, economical, and user-dependent issues to be resolved. Nanotechnologies have influenced the development of fuel cell technologies in many ways. Specifically portable-device-friendly technologies, like polymer electrolyte membrane fuel cells (PEMFC), are currently under research for low-temperature fuel cell development, high-volume processing methods for membrane–electrode assemblies, and effective dispersion of expensive metal catalysts [18]. An important factor for energy density, the oxygen reduction reaction, currently a limiting factor in low-temperature fuel cells, depends on the size, organization, and shape of platinum nanoparticles, a few nanometers in size [19]. Platinum nanoparticles are studied using second metal surface reactions to reduce the direct methanol fuel cells (DMFC) limitations, caused by the use of methanol and CO/CO_2. When oxidization is the wanted reaction, CO/CO_2 reduces the effective surface area of the active catalyst particles. Nanotechnologies improve industrial scale production by introducing efficient monolayer assemblies of novel catalyst nanoparticles with high surface reactivity.

Polymer electrolyte membranes, like Nafion^{TM}, are used in many of the studied assemblies, both in PEMFCs and DMFCs. Fuel cells are classified according to the type of electrolyte. The most common electrolyte membranes, such as Nafion, are permeable to protons. The second most common electrolyte membranes, found in solid oxide fuel cells (SOFCs), are permeable to oxide ions. A third type of electrolyte, used for molten carbonate fuel cells at high temperatures conducts carbide ions (CO_3^{2-}). Alkaline solutions have also been used in some of the fuel cells. Currently polymer membranes are only capable of low-temperature processes, and thus new materials, like block copolymers have been studied for ion channel formation. Nanostructured membranes have been shown to enhance proton conduction and mechanical stability. For practical development of fuel cells, the most promising source, hydrogen, has to be stored efficiently to increase the time between tank fillings. Nanomaterials promise efficient hydrogen storage with CNTs and metal–carbon–hybride composites [20, 21]. High-temperature cells (solid oxide and molten carbonate electrolytes) operate by very different mechanisms than low-temperature cells, and have different applications accordingly.

Fuel cell technology has promised much, but several technical and nontechnical issues still remain to be solved:

- miniaturization for portable devices remains as a challenge in fuel cell development for the fuel delivery, micro-fluidics and power electronics parts;
- a fuel cell recharger may not be capable of achieving better energy density than conventional electro-chemical rechargers;
- commercial distribution channels for the fuel have not yet been established and there are issues concerning its safe transportation;
- use of fuel cells would need a change in user behavior – carrying and filling a recharger. Customer/user habits are sometimes not easy to change, when establishing new ways of working.

In summary, it can be argued that fuel cells may eventually provide flexible energy solutions for consumer electronics. There are, however, still several steps between the current state-of-the-art development and an approved, safe, and user-friendly solution. It is likely that fuel cells, predominantly PEMFC and DMFC, will emerge in the military portable devices and emergency device markets, where an instant source of energy is needed and cost is not an issue.

3.3   Energy harvesting – nanotechnologies in portable systems

The development of electonics outpaced the development of energy storage systems a long time ago. It is no longer possible to miniaturize a modern mobile phone just by decreasing the size of the electronics and electronics packaging. Although much of the real estate on a circuit board is taken by integrated circuits and currently also by passive circuit elements, like resistors, bypass capacitors, and circuit protection devices, it would not make much difference to the size of the whole system if the electronics real estate were to be halved. The controlling factors in mobile devices are the mechanical parts, user interface, and energy source. For device slimness, battery and display size play a key role, while other mechanical parts take roughly one third of the body volume.

There is no doubt that nanotechnologies will have a great influence in how low-power electronics will develop in the future. The ability to exploit the remarkable properties of CNTs may be the first step in this process. CNTs have already been used successfully in some novel experimental electronic devices, transistors, and various interconnect wires, via contact structures in integrated circuits. The high electrical and thermal conductivity properties of nanotubes, together with their structural rigidity at the nanoscale are exploited in such applications.

Nanotechnologies can provide enhanced solutions for portable devices and sensor nodes. Firstly, in energy storage, battery and capacitor electrode development for higher EDLC layer capacitance requires a very thin layer of double-layer capacitance, which is highly dependent on the specific surface area of the structure and micro- and nanosize pores. Lithium ion battery anodes, like supercapacitor electrodes, have been the focus of interest of many research groups in looking for the right kind of nanoporous carbon for the ultimate energy storage solution; a battery with low ESR high-power output. Secondly, future energy delivery and packaging solutions will draw on nanostructures and materials developed for high-power cables rated to the requirements of large power grids. On a small scale, higher conductivity solutions for power delivery networks inside miniaturized devices can be used to build stronger, more robust, and low-power solutions, with enhanced heat dissipation and heat tolerance properties. Conductivity and wear resistance can be achieved with CNT enhanced contact surfaces and CNT heat sinks contacts. Thirdly, nanotechnology could act as an enabler for new energy scavenging solutions for portable electronics devices and autonomous devices.

3.3.1   Energy from temperature gradients

Waste heat collection has some potential as an energy source for small devices [22]. Theoretical maximum efficiency, Carnot efficiency, depends on the temperature difference, $\eta = (T - T_0)/T_0$. However, for portable devices, or devices in the natural environment temperature gradients are small: 5 °C or 10 °C differencies could be achieved. For a 5 °C temperature difference the maximum efficiency would be 1.6%. Applications for heat scavenging would be more practical in environments where heat is generated at a high temperature relative to the ambient, e.g., in high-power electronics (network server processors, power converters), vehicular systems (cars, planes, and space applications) or space systems, where the temperature differences can be large. Heat energy scavenging modules have been developed by companies such as HiZ Technology, Kryotherm, and Tellurex for 300–500 mW powers in sizes of few cubic centimeters. However, the temperature differences are relatively large, $\Delta T = 80\,°C$. Nanotechnologies could come into play in allowing control of the thermal conductivity of the crystal lattices, the use of quantum confinement effects with quantum dots to design energy flow, and charged particle exchange between the electrodes.

3.3.2   Ambient RF and vibration energy – micro- and nanowatt level power harvesting

Ambient energy harvesting through collection of RF energy will become viable in the future, as more networks with denser transmitter networks at different frequencies are deployed. Presently a wide range of WLAN frequencies, WLAN 2.5/5 GHz, cellular frequencies, GSM/CDMA 400 MHz, 800 MHz–2.1 GHz, are in use. Novel wideband radios, such as Wimax/UWB, GPS, DVB-H, and FM radios integrated in current media-rich mobile phones, will not produce adequate energy density for power harvesting applications, because of the large distance between transmitters and receivers. The maximum transmitter power delivered can be calculated $P = P_0\lambda^2/4\pi R^2$, which shows that at the optimum it varies as the square of the distance. With a 1 W transmitter (limited for inside/office use by federal regulations in the USA) power received at 5 m distance would be 50 µW at most [23].

Electromagnetic energy harvesting with very low power levels is limited by the low efficiency of power conversion. Conversion of very low voltage levels to the higher voltage levels applicable to charging a storage device or to being used in processing or communication is problematic, because the efficiency of the conversion is reduced with the lower levels of power and voltage. This is due to losses in the first element of the voltage conversion in the active power conversion using charge pump circuits. Passive elements can also be used, but typically proposed solutions at higher energy levels use inductors, which are not suitable at very low energy levels. Passive diode–capacitor networks could be a solution. In diode–capacitor network solutions, and in active-switching-based solutions using diodes, the diode threshold voltage and forward drop become crucial parameters. The lowest voltage thresholds can be found in Schottky diodes, where a 200 mV turn on voltage levels can be achieved. However, this still implies that the energy density in the RF field must be high enough to achieve a voltage of 200 mV at the rectifier input.

Some novel methods for voltage amplification of RF energy are under investigation. Efficiencies of 1–20% harvesting of very low energy levels (10^{-5}–10^{-1} mW) have also been proposed and demonstrated for wideband 1–10 GHz, using Schottky diodes and specially constructed antenna arrays with self-matching structures.

In indoor environments it is feasible to harvest and thus recycle multiband, multipolarized, and spatially distributed RF energy to power distributed sensor networks [23]. In small portable devices large antenna array elements are not applicable.

Current mobile phones typically have planar inverted F antennas (PIFAs), optimized for a few watts of RF transmission and milliwatt level cellular frequency data reception. Front-end solutions incorporate switches for selecting the radio receiver/transmitter direction and typically all cellular terminals incorporate surface acoustic wave (SAW) filters for analog and efficient pre-processing of the wave. SAW devices implement DSP-type processing in the analog signal phase. Depending on the dimensions, the frequency response of the device can be modified.

A well-known property of SAW devices is energy storage. Based on the piezoelectric functionality of the propagation of the wave in the SAW device, they can be used as receivers as well as analog signal processors. SAW devices are comparatively small and could be used as receiving antennas in future terminals. An array of roughly 1 mm^2 SAW devices can be used to convert very-low-power ambient RF signals to much higher voltage signals. This is enabled by the high quality factor (Q) of the SAW resonator. The meaningful size of such an array of SAW devices for energy harvesting can be as small as a typical PIFA antenna in a mobile phone. Wideband SAW devices would be needed to be able to receive broad bands of RF energy. These devices would have to be

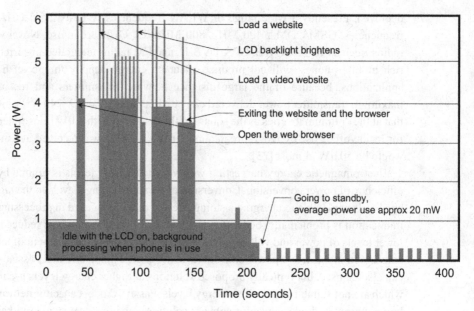

Figure 3.9 Typical pattern of power consumption of a mobile phone used for web browsing and mobile video rendering.

connected to low-voltage energy conversion systems for efficient energy storage. Power levels could be designed to be in the range of the portable internet terminal quiescent power levels, 2–18 mW, or the levels of autonomic sensor device power levels 1–2 mW at minimum.

An average modern mobile phone in active use can consume up to 2 W of power. Figure 3.9 gives an example of the power consumption of a mobile phone as a function of time over a period of 300 seconds. Peaks of over 6 W (Figure 3.9) of power in communication mode (browsing internet, video calls, local radio interfacing) are not exceptional. The greatest power is used when searching the network (especially in multinetworks) and during high-speed transmission. Nevertheless, one third of the overall energy is used in the standby mode – when the phone is idle and waiting for the communication to start-up. The amount of power used in standby mode is on average 20 mW, of which the major portion is consumed by the radio components (analog to digital and digital to analog converters, RF receiver voltage reference and DC/DC converter for power amplifier).

RF harvesting technologies capable of harvesting 10–30 mW of energy could in this context enhance the standby time of a device markedly. A device with the current typical standby time of 5 days would be capable of receiving calls after 10 days, if it could constantly harvest an average of 10 mW of waste electromagnetic energy from its surroundings. If more energy were available, the device could even charge itself, ultimately making the wall plug charger necessary only as back up, i.e., it would be an energy autonomous device. For such low power levels, however, the charging time would be inconveniently long – days or weeks. However, for emergency device use,

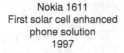

Nokia 1611
First solar cell enhanced
phone solution
1997

Solar cells appear in chargers,
toys, and low-power accessories,
such as Bluetooth headsets

Nokia Morph
nanotechnology-based
device concept
with ambient energy harvesting

1995   2005   2015

Figure 3.10 Evolution of solar cell concepts in mobile devices.

when no AC mains power is available, in isolated locations or for very-low-power devices (autonomous sensors, tags) such low levels could be useful.

The technical bottleneck is the lack of ultralow-power charging circuits and of a convenient way of using very efficient and wideband antennas or antenna arrays. Here nanotechnologies could be used. CNTs have been studied for use in radio communication with tuning matched to the CNT natural mechanical vibration (a function of tube length and diameter) with the rectification function being achieved by biasing the CNT as a field emitter in the diode configuration. Audio reception has been demonstrated. However, for energy harvesting this method needs an external power source, with harvested energy being energy added to the supply. Nanowires have been studied for harvesting wideband electromagnetic energy, and been proposed for use in antenna receivers. The current limitations are the low efficiency of CNTs as antennas and their inductive behavior.

The special properties of CNTs as transmission lines, on the other hand, suggest that they can also be used as ultracompact resonant antennas [24]. RF vibrations can also be induced in micro- or nanomechanical structures that have piezoelectric properties. Current could be amplified using a forest of vertically grown nanowires, or micromechanical cantilevers. Such devices are still at an early stage of research [25–27].

3.3.3   Portable solar cells and nanotechnologies

3.3.3.1   P–n junction solar cells with nanowires or carbon nanotubes

The use of solar cells to power mobile phones was first demonstrated in 1997, as depicted in Figure 3.10. These early trials and the resulting development have not yet resulted in any significant consumer applications. This has been mostly due to cost and to the limited surface area of mobile devices. However, solar cells have been used in various handheld devices, toys, and chargers for mobile devices. New lower-cost materials and photovoltaic devices based on nanotechnology may enable new solar energy solutions for mobile devices, as illustrated in the Nokia Morph concept (see

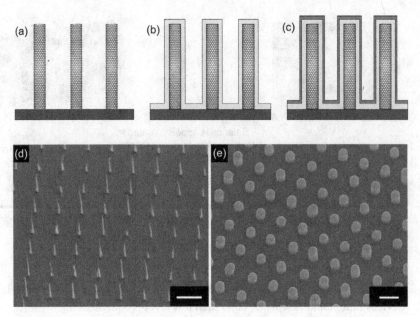

Figure 3.11 Schematics of device fabrication and SEM images of coaxial MWNT–amorphous silicon solar cell arrays: (a) growth of patterned MWNTs on tungsten; (b) deposition of amorphous silicon onto MWNTs arrays; (c) deposition of transparent conductive top electrode; (d) SEM image of the as-grown MWNTs array; (e) SEM image of the amorphous Si coated MWNTs array. Scale bars are 2 μm.

Figure 3.10). Various nanotechnologies have been explored with the aim of boosting the efficiency of photovoltaic devices. For example, inorganic nanoparticles and nanowires have been incorporated into organic solar cells, in order to extend the interface for charge separation and improve carrier mobility. For conventional solar cells made of inorganic materials, it was proposed that nanowire-like structures could potentially lead to a better collection efficiency of photo-generated electron–hole pairs together with improved optical absorption. Particular emphasis has been placed on the coaxial core–shell nanowire structure, where charge separation occurs in the radial direction, i.e. orthogonally to the carrier transportation path. This idea can be implemented either with a core–shell nanowire p–n junction or with a core–shell structure formed by two materials with type-II band alignment, which separate charge at the interface without the need for doping. From a practical point of view, however, clear evidence is missing about nanowire solar cells outperforming existing solutions.

Silicon-based solar cells currently dominate the photovoltaic (PV) market. Solar cells based on coaxial silicon nanowires (SiNWs) have been fabricated, while the power conversion efficiency of SiNW p–n and p–i–n junctions has been measured down to the individual nanowire level. The interpenetrating p–n junction geometry based on coaxial nanowires, however, still relies on carrier collection paths made of semiconductor materials, which pose constraints on collection efficiency due to path resistance.

Figure 3.11 depicts a core–multishell nanowire structure used to realize solar cells with interpenetrated metallic electrodes [28]. This approach was based on coating

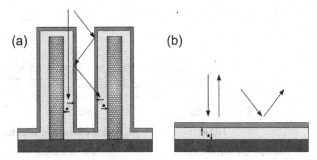

Figure 3.12 Schematic diagrams that compare the light-trapping structure and the charge carrier collection direction of: (a) MWNT-based solar cells; and (b) planar conventional solar cells.

vertically aligned MWNTs with amorphous silicon (a-Si:H) shells and indium tin oxide (ITO). The use of MWNTs as core contacts can avoid electrical losses that might occur in other types of nanowires. Moreover, nanotube/nanowire arrays form a natural light-trapping structure. Indeed, a 25% increase of the short-circuit current was achieved for the nanotube/nanowire array compared to the planar a-Si cell used as the reference.

The efficiency of a-Si:H thin film solar cells can be enhanced by effective light trapping, especially in the red spectral band where absorption is weak. Figure 3.12 illustrates the light-trapping model implemented in the array of coaxial MWNTs/a-Si:H solar cells. The coaxial solar cell arrays depicted in Figure 3.12(a) offer two distinct optical advantages over a conventional thin-film structure. First, planar cells with no light-trapping capabilities require an antireflection coating to reduce reflection and achieve good optical confinement. This solution, however, only ensures antireflection over a narrow spectrum. Second, one must consider that, for a planar solar cell structure (Figure 3.12(b)), there will be more charge carriers generated on the illuminated side of the cell when the light hits the surface. One species of the photo-generated carriers will therefore have a longer path to travel than the other before collection by the electrode. Conversely, in the radial junction structure (Figure 3.12(a)), while light is being absorbed along the axial direction, both electrons and holes are moving radially, and on average cover an equal distance to reach the metallic junction before charge collection. The final outcome is that the average carrier drift time of a particular type is shortened, and the chance for electron–hole recombination in the absorption region is reduced, leading to a better diode ideality factor.

3.3.3.2 Dye-sensitized solar cells

Photoelectrochemistry, in which light is converted into electricity with efficiencies comparable to those achievable with inorganic semiconductor photovoltaic cells, has attracted much attention.

A typical photocurrent-generating electrochemical device has a semiconductor in contact with an electrolyte, and is often referred to as a photoelectrochemical cell. A photoelectrochemical cell consists of a photoactive semiconductor working electrode (either n- or p-type) and a counterelectrode made of either metal (e.g., Pt) or

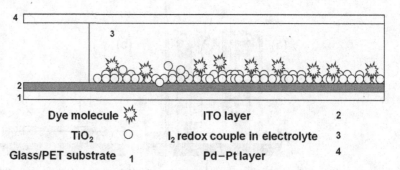

Figure 3.13 Structure of a dye-sensitized solar cell (DSSC).

semiconductors. Both electrodes are immersed in an electrolyte containing suitable redox couples. In a metal–electrolyte junction, the potential drop occurs entirely on the solution side, whereas in a semiconductor–electrolyte junction, the potential drop occurs on the semiconductor side as well as the solution site. The charge on the semiconductor side is distributed deep in the interior of the semiconductor, creating a space charge region. If the junction of the semiconductor–electrolyte is illuminated with a light having energy greater than the bandgap of the semiconductor, photo-generated electrons and holes are separated in the space charge region. The photo-generated minority carriers arrive at the interface of the semiconductor–electrolyte. Photo-generated majority carriers accumulate at the back of the semiconductor. With the help of a connecting wire, photo-generated majority carriers are transported via a load to the counterelectrode where these carriers electrochemically react with the redox electrolyte.

A pioneering photoelectrochemical experiment was the measurement of the photocurrent between two platinum electrodes immersed in an electrolyte containing metal halide salts. It was later found that the photosensitivity can be extended to longer wavelengths by adding a dye to silver halide emulsions. In this case the photoabsorption occurs within the halide-containing electrolyte. Interest in photoelectrochemistry of semiconductors led to the discovery of wet-type photoelectrochemical solar cells. These studies showed electron transfer to be the prevalent mechanism for the photoelectrochemical sensitization processes. Grätzel has extended the concept to dye-sensitized solar cells (DSSCs).

The first promising photoelectrochemical solar cell was based on a dye-sensitized porous nanocrystalline TiO_2 photoanode. Interest in porous semiconductor matrices permeated by an electrolyte solution containing dye and redox couples was greatly stimulated by this and similar cells. The conversion efficiency of the DSSC has been improved to above 11% since the first DSSC reported had an efficiency of 7.1% [29]. Even though the best crystalline silicon cells have attained ∼25% efficiency, the maximum theoretical conversion efficiency is approximately 30% for both devices. Large DSSCs have been prepared by screen printing an embedded silver grid on a fluorine-doped tin oxide (FTO) glass substrate. Under standard test conditions, the energy conversion efficiency of the active area was 5.52% in a 5 cm × 5 cm device. The efficiency of a small equivalent was 6.16%. In a DSSC, the initial photoexcitation occurs in the light absorbing dye (Figure 3.13). Subsequent injection of an electron from the photo-excited

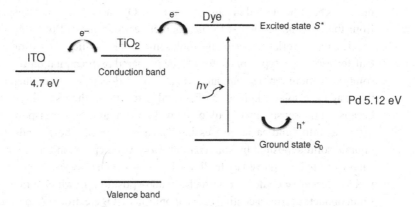

Figure 3.14 Energy diagram of a DSSC.

dye into the conduction band of semiconductors results in the flow of current in the external circuit (Figure 3.14). Sustained conversion of light energy is facilitated by regeneration of the reduced dye either via a reversible redox couple, which is usually I_3^-/I^- or via an electron donation from a p-type semiconductor mixed with the dye.

The necessity to absorb far more of the incident light in a DSSC was the driving force for the development of mesoscopic semiconductor materials with an enormous internal surface area. High photon to electron conversion efficiencies were reported based on the dye-sensitized mesoporous TiO_2 solar cells. The major breakthrough in DSSCs was the use of a high-surface-area nanoporous TiO_2 layer. A single monolayer of the dye on the semiconductor surface was sufficient to absorb essentially all the incident light in a reasonable thickness (several micrometers) of the semiconductor film. TiO_2 has become the semiconductor of choice, since it is cheap, abundant, and nontoxic.

The standard dye used is tris(2,2'-bipyridyl-4,4'-carboxylate)ruthenium (N3 dye). The function of the carboxylate group in the dye is to attach to the semiconductor oxide by chemisorption. The dye must carry attachment groups such as carboxylate or phosphonate to graft itself firmly onto the TiO_2 surface [30]. The attachment group of the dye ensures that it spontaneously assembles as a molecular layer upon exposing the oxide film to the dye solution. Once a photon is absorbed, there is therefore a high probability that the excited state of the dye molecule will relax by electron injection to the semiconductor conduction band. The dye N3 was discovered in 1993 and for some time dominated the field, although a credible challenger has now been identified; tri(cyanato-2,2',2''-terpyridyl-4,4',4''-tricarboxylate) Ru (black dye). The response of the black dye extends 100 nm further into the infrared area of the spectrum than that of the N3 dye.

Because of the encapsulation problem posed by use of liquid in the conventional wet-type DSSC, much work is being done to make an all-solid-state DSSC. The use of solvent-free electrolytes in the DSSC should in principle offer very stable performance. To construct a fully solid-state DSSC, a solid p-type conductor needs to be chosen to replace the liquid electrolyte. The redox levels of the dye and p-type materials have to be adapted carefully. The result is an electron in the conduction band of n-type semiconductors (e.g. TiO_2) and a hole localized in the p-type conductor. Hole transporting amorphous materials have been used in nanocrystalline TiO_2-based DSSC to transport hole carriers from the dye cation radical to the counterelectrode instead of using the I_3^-/I^- redox species. Early work focused on the replacement of I_3^-/I^- liquid electrolyte with CuI [31]. CuI for use as a p-type conductor can be prepared by precipitation from an acetonitrile solution at room temperature and it is also a solid-state ionic conductor. Cells made in this way have solar efficiencies of several percent, but their stability is relatively poor, because of the sensitivity of CuI to air and light. Organic hole transporting materials will offer flexibility and easier processing. Bach *et al.* used a hole conducting amorphous organic solid deposited by spin coating. However, deposition on nanoporous materials cannot easily be achieved by traditional methods such as evaporation or spin coating. A viable alternative could be electrochemical deposition, in which a thin layer of organic semiconductor forms a coating on a nanoporous TiO_2 electrode. A low bandgap polymer consisting of layers of thiophene and benzothiadiazole derivatives has been used in a bulk heterojunction DSSC. This solid-state DSSC exhibits power conversion efficiency of 3.1%, which is the highest power conversion efficiency achieved with organic hole transporting materials in a DSSC to date.

Construction of quasi-solid-state DSSCs has also been explored. Quasi-solid-state DSSCs are based on a stable polymer grafted nanoparticle composite electrolyte, cyanoacrylate electrolyte matrix, and a novel efficient absorbent for the liquid electrolyte based on poly(acrylic acid)–poly(ethyleneglycol). The polymer gels in the above cases function as ionic conductors. The melting salts, also referred to as room temperature ionic liquids, are also good ionic conductors. Solid-state DSSCs based on ionic liquids can be made to enhance the conversion efficiency. These devices use imidazolium-type ionic liquid crystal systems as the effective electrolyte. The use of ionic liquid oligomers (prepared by incorporating imidazole ionic liquid with polyethylene oxide oligomers) as the electrolyte in a DSSC, shows that the increase of the polyethylene oxide molecular weight in the ionic liquid oligomers results in faster dye regeneration and lower charge transfer resistance of I_3^- reduction. This leads to an improvement in the DSSC performance.

However, the main limiting factors in a DSSC based on ionic liquids compared with the conventional wet-type DSSC are the higher recombination and lower injection of charge. At low temperatures, the higher diffusion resistance in the ionic liquid may also be the main limiting factor through its effect on the fill factor.

Plastic and solid state DSSCs which incorporat single-walled carbon nanotubes (SWCNTs) and an imidazorium iodide derivative have been fabricated. The introduction of CNTs improves the solar cell performance through reduction of the series resistance. TiO_2 coated CNTs have been used in DSSCs. Compared with a conventional TiO_2 cell, the TiO_2-CNT (0.1 wt%) cell gives an increase in short circuit current density (J_{SC}), which can lead to an increase of the conversion efficiency of $\sim$50%, from 3.32% to 4.97%. It is thought that the enhancement of J_{SC} is due to improvement in interconnectivity between the TiO_2 particles and the TiO_2-CNTs in the porous TiO_2 film. When employing SWCNTs as conducting scaffolds in a TiO_2-based DSSC, the photoconversion efficiency can be boosted by a factor of 2. The internal photocurrent efficiency (IPCE) is enhanced significantly to 16% when the SWCNT scaffolds support the TiO_2 pariticles. TiO_2 nanoparticles dispersed on SWCNT films improve photo-induced charge separation and transport of carriers to the collecting electrode surface. Besides TiO_2 nanoparticles, ZnO nanoparticles and nanowires have also been used for low-temperature processing of DSSCs on flexible substrates [32, 33].

In comparison with the expensive crystalline silicon p–n-type solar cells, DSSCs show a lower photovoltaic to electric power conversion efficiency, but have an efficiency comparable to amorphous silicons. Their advantages lie in versatility of the material applications and the possibility to apply a wider range of adsorbent materials and different (nano)materials with flexible and transparent structures. DSSCs are suitable for energy harvesting, where a wide spectrum of low-intensity light, including artificial light, is photovoltaically converted to provide auxiliary power. Given its low cost of production, they could be deployed over large areas in a ubiquitous manner. For example, when coupled with a battery/supercapacitor, they could act as a renewable energy point for charging mobile devices in buildings. Silicon cells which are used for primary power generation from the sun are not cost-efficient for deployment in low-light-intensity environments.

3.4   Conclusions

The energy and power bottleneck is a major hurdle to overcome in increasing the functionality of mobile devices while also maintaining a small form factor and low weight, which are overriding considerations when deciding whether a device is "mobile." One of the developments used to address this is the portioning of the mobile device power system in such a way that the peak power requirement is not met directly through the battery but rather through a supercapacitor. This enables the battery to be optimized purely for energy density, without its sizing also being limited by peak current (power) capability. The role of nanotechnologies in enhancing lithium ion battery, fuel cell, and supercapacitor performance have been reviewed. In the main this is through nanoscale structuring of electrodes, which allows enhancement of energy/power capacity through increase in specific surface area, lowering of ESR, and enhanced surface reactivity. Enhancing surface reactivity through exploitation of nanoscale surface chemistries designed for specific ionic species is at a very early stage of research. This promises to be an avenue that can revolutionize the expected performance from chemically based energy sources such as batteries, fuel cells, and electrochemical double layer supercapacitors. A promising form of nanoscale carbon, which is being evaluated for enhancing surface area as well as reactivity, is the CNH.

As well as enhancement in the capacities of electrochemical cells and supercapacitors, future mobile devices will also rely on energy harvesting from the environment. The aim here is limited to complementing the main energy/power system of the mobile device based on a battery and supercapacitors. For example, if the energy capacity in the mobile device can be enhanced by 10–20% without a corresponding increase in battery capacity, a significant enhancement in functionality such as increased video upload time can be achieved at "no energy cost." Having supercapacitors in the mobile energy system is particularly suited for energy harvesting. This is because most environmentally harvestable energy sources are intermittent and transient in nature. Therefore being able to capture energy in pulses of power is a key requirement for future mobile energy systems. This can be achieved through the use of supercapacitors that are particularly suitable for rapid pulse charging and storage of energy.

Nanotechnologies will also be key for energy harvesting devices, which are suitable for incorporation within a mobile platform. The form factor and weight of these devices have to be compatible with those of the platform. Emerging photovoltaic devices which exploit nanoscale materials for performance enhancement and form factor flexibility have been reviewed. Energy harvesting from the RF field through the use of new wideband antenna structures and intermediate converters such as SAW devices is also a viable route for the future. CNT antenna structures that are particularly suitable for this purpose are in development.

Taken as a whole, the energy and power systems for mobile devices are poised to become a subdiscipline in their own right. They encompass many exciting and new technologies that have only become viable through the advent of engineering and science at the nanoscale. In turn nanotechnologies that are focused primarily on the requirements of mobile energy/power systems are a rapidly expanding area of research.

References

[1] *Future Mobile Handsets*, 8th edition, Worldwide Market Analysis & Strategic Outlook, pp. 2006–2011, Informa UK Ltd, 2006.

[2] B. E. Conway, *Electrochemical supercapacitors – Scientific Fundamentals and Technological Applications*. Kluwer Academic, 1999.

[3] M. Rouvala, Characterization of carbon nanostructure electrode surface for low ESR battery structures. Dissertation for Master of Philosophy, University of Cambridge, 2008.

[4] A. S. Aricò, P. Bruce, B. Scrosati, J.-M. Tarascon, and W. van Schalkwijk, Nanostructured materials for advanced energy conversion and storage devices. *Nat. Mat.*, **4**, 366–377, 2005.

[5] D. Larcher, C. Masquelier, D. Bonnin, *et al.*, Effect of particle size on lithium intercalation into α-Fe_2O_3. *J. Electrochem. Soc.*, **150**, A133–A139, 2003.

[6] T. D. Burchell, *Carbon Materials for Advanced Technologies*, pp. 341–385, Pergamon, 1999.

[7] D. Linden and T. B. Reddy, *Handbook of Batteries*, third edition, pp. 35.1–35.21, McGraw-Hill, 2001.

[8] S. Bourderau, T. Brousse, and D. M. Schleich, Amorphous silicon as a possible anode material for Li-ion batteries, *J. Power Sources*, **81–82**, 233–236, 1999.

[9] J. R. Dahn and A. M. Wilson, Lithium insertion in carbons containing nanodispersed silicon. *J. Electrochem. Soc.*, **142**, 326–332, 1995.

[10] C. Chan, H. Peng, G. Liu, *et al.*, High-performance lithium battery anodes using silicon nanowires, *Nature Nanotechnology*, **3**, 31–35, 2008.

[11] Y. Yudasaka, S. Iijima, and V. H. Crespi, Single wall carbon nanohorns and nanocones in *Carbon Nanotubes*, G. Dresselhaus, M. S. Dresselhaus, and A. Jorio, eds., Topics Appl. Physics 111, Springer-Verlag, 2008.

[12] H. Wang, M. Chhowalla, N. Sano, S. Jia and G. A. J. Amaratunga, Large-scale synthesis of single-walled carbon nanohorns by submerged arc. *Nanotech.*, **15**, 546–550, 2004.

[13] Y. Gogotsi, *Nanomaterials Handbook*, CRC Press, 2006.

[14] K. An, W. Kim, Y. Park, *et al.*, Supercapacitors using single-walled carbon nanotube electrodes. *Adv. Mater.*, **13**, 1425–1430, 2001.

[15] C. Du and N. Pan, High power density supercapacitor electrodes of carbon nanotube films by electrophoretic deposition, *Nanotech.*, **17**, 5314–5318, 2006.

[16] C. Du, J. Yeh, and N. Pan, High power density supercapacitors using locally aligned carbon nanotube electrodes, *Nanotech.*, **16**, 350–353, 2005.

[17] J. Chimiola, G. Yushin, Y. Gogotsi, C. Portet, P. Simon, and L. Taberna, Anomalous increase in carbon capacitance at pore sizes less than 1 nanometer, *Science*, **313**, 1760–1763, 2006.

[18] C. Wang, M. Waje, X. Wang, J. M. Tang, R. C. Haddon, and Y. Yan, Proton exchange membrane fuel cells with carbon nanotube based electrodes, *Nano Lett.*, **4**, 345–348, 2004.

[19] X. Wang, M. Waje, and Y. Yan, CNT-based electrodes with high efficiency for proton exchange membrane fuel cells, *Electrochem. Solid State Lett.*, **8**, A42–A44, 2004.

[20] M. Waje, X. Wang, W. Li, and Y. Yan, Deposition of platinum nanoparticles on organic functionalized carbon nanotubes grown in situ on a carbon paper for fuel cells. *Nanotech.*, **16**, S395–S400, 2005.

[21] Z. Chen, L. Xu, W. Li, M. Waje, and Y. Yan, Polyaniline nanofibers supported platinum nanoelectrocatalysts for direct methanol fuel cells, *Nanotechnology*, **17**, 5254–5259, 2006.

[22] M. T. Penella, and M. Gasulla, Warsaw, a review of commercial energy harvesters for autonomous sensors, in *Proceedings of the 2007 IEEE Instrumentation & Measurement Technology Confrence, IMTC 2007,* R. Thorn, ed., IEEE 2007.

[23] J. A. Hagerty, F. B. Helmbrech, W. H. McCalpin, R. Zane, and Z. B. Popovic, Recycling ambient microwave energy with broad-band rectenna arrays. *IEEE Trans. Microwave Theory And Techniques*, **52**, 1014–1024, 2004.

[24] C. Rutherglen and P, Burke, Carbon nanotube radio. *Nano Lett.*, **7**, 3296–3299, 2007.

[25] H. Bergveld, W. Kruijt, and P. Notten, *Battery Management Systems – Design by Modeling*, Philips research book series, vol. 1, pp. 50–52, Kluwer Academic Publishers, 2002.

[26] A. S. Aricò, P. Bruce, B. Scrosati, J.-M. Tarascon, and W. van Schalkwijk, Nanostructured materials for advanced energy conversion and storage devices, *Nat. Mat.*, **4**, 253–260, 2005.

[27] B. E. White, Energy-harvesting devices: beyond the battery, *Nat. Nanotech.*, **3**, 71–72, 2008.

[28] H. Zhou, A. Colli, Y. Yang, *et al.*, Arrays of coaxial multiwalled carbon nanotube-amorphous silicon solar cells, *Adv. Mater.* to be published.

[29] M. Grätzel, Photoelectrochemical cells, *Nature*, **414**, 338–344, 2001.

[30] B. O'Regan and M. Grätzel, A low-cost, high-efficiency solar cell based on dye-sensitized colloidal TiO_2 films, *Nature*, **353**, 737–740, 1991.

[31] U. Bach, D. Lupo, P. Comte, *et al.*, Solid-state dye-sensitized mesoporous TiO_2 solar cells with high photon-to-electron conversion efficiencies, *Nature*, **395**, 583–585, 1998.

[32] D. Wei, H. E. Unalan, D. Han, *et al.*, Solid-state dye-sensitized solar cell based on a novel ionic liquid gel and ZnO nanoparticles on a flexible substrate, *Nanotech.*, **19**, 424006, 2008.

[33] H. E. Unalan, D. Wei, K. Suzuki, *et al.*, Photoelectrochemical cell using dye sensitized zinc oxide nanowires grown on carbon fibers, *Appl. Phys. Lett.*, **93**, 133116, 2008.

4 Computing and information storage solutions

P. Pasanen, M. A. Uusitalo, V. Ermolov, J. Kivioja, and C. Gamrat

4.1 Introduction

The mobile devices of the future are expected to be able to communicate with each other wirelessly at ever increasing data rates, and to be able to run a vast number of applications, with a great need for more computational speed and power. Indeed, it is foreseeable that approaching data rates of gigabits per second can alone be a challenge: large bandwidths in the gigahertz range and increased complexity in interference cancelation and error correction coding, combined with cognitive radio and multiple antenna techniques could lead to computing and power consumption needs that are extremely challenging with current conventional methods or any of their expected evolutions. To combat this, there has been a resurgence of interest in application-specific processing, instead of general all-purpose processors. One suggestion for possible future mobile phone architecture is based on the so-called network-on-terminal architecture (NoTA) [74], pictured in Figure 4.1, where different subsystems can be connected via standardized interconnects. Each subsystem consists of computing and memory elements, targeted at specific applications. This will allow more freedom in the design of the subsystem processors, and therefore ease the introduction of novel computing technologies into mobile phones, perhaps some of which could be based on new nanotechnology-enabled computing elements.

Nanocomputing has been attracting a lot of attention since the arrival of efficient tools for nanoscale manipulation of matter. The hope is that the new opportunities provided by nanotechnology will provide faster computation and signal processing systems so that challenging computational problems can be solved more efficiently, possibly using new computational principles. Besides the raw computational power, there are other very interesting opportunities for nanocomputing systems – ambient intelligence being one of the most intriguing. The computational elements realized by nanotechnology could become smaller, more energy-efficient, cheaper, easier to integrate into different materials, and more environmentally friendly. Devices could be embedded almost everywhere; they could operate even without charging and be able to understand their environment, in order to provide more seamless interaction with

Nanotechnologies for Future Mobile Devices, eds. T. Ryhänen, M. A. Uusitalo, O. Ikkala, and A. Kärkkäinen.
Published by Cambridge University Press. © Cambridge University Press 2010.

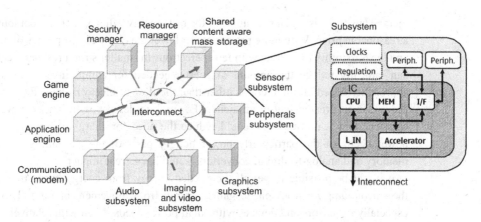

Figure 4.1 Example of a possible multiprocessor architecture (NoTA) for future mobile phone.

humans. This will require a high level of autonomous operation in terms of power management, communication, and especially computing and information processing skills. In these applications, the computing speed is not necessarily the important figure of merit. For instance, due to massively parallel computation or highly application-specific architectures even slow operation of individual units could yield enormous collective information processing capabilities.

Nanotechnology could enable selective and sensitive detection of a broad range of measurands. For example, miniaturized sensors for detecting biological and chemical substances, light or other electromagnetic radiation, charge, physical touch as well as small weights and pressure (including audio) could be realized much more cost effectively with nanoelements. The large number of nanosensors will generate a lot of data: it will be essential to integrate the computing with the sensing and to process the data before feeding it forward to ensure fast and energy-efficient signal processing. Only meaningful and classified data would be transferred forward and only when it is necessary (triggering). This would obviously require a very low cost of fabrication: devices should be affordable to purchase and operate.

Ease of use and invisibility of the devices are also required: most people do not want to have to learn how to use new technology, instead they want the interfaces to services and applications to be so intuitive and natural that using them does not require additional effort or learning. Here, nanotechnology could enable new forms of user interfaces and integration of the functionalities into the matter around us. The devices need to have better capabilities for recognition and understanding of speech, text, and images, as well as robotic control. These currently challenging applications might be realized by an efficient implementation of nanotechnology-based machine learning and other algorithms, developed specifically for these purposes.

Energy efficiency is of paramount interest for any mobile device or sensor. Future ambient intelligence implies an increasing number of devices, each consuming energy over their lifetime. Mobile devices have only a limited capacity to carry energy with them. Thus there is an increasing need for fast, energy-efficient general computation at a reasonable price. With more precise control over the electrical properties of materials, and the possibility to operate with fewer electrons in smaller structures, nanotechnology can radically reduce the amount of energy needed for computing tasks. Moreover, approaches based on interactions other than the flow of electrons have the potential to provide even greater energy efficiency. For mobile devices with higher power needs, nanotechnological computing could reduce the need for charging. Here reducing the power needs of the particularly power hungry parts of the computing system, like memory and analog-to-digital converters, could have great impact.

It might be possible to get all the needed energy from energy harvesting, i.e., use the surrounding environment to gather energy from movement or light. This applies especially to sensors and devices with small power needs. Even with relatively efficient energy harvesting this would require extremely low power consumption in computing. For example, with magnetic nanoparticles one might be able to realize a computing system that would stop when there is a lack of energy and continue when new energy is available.

In many cases heat management is a limiting factor in efficient computing. Often the capability to store enough energy is less critical than the ability to remove the heat generated by the components. As discussed in Chapter 3, handheld devices of current dimensions and form factors can stand 1–6 W of energy consumption while keeping their surface temperatures within safe limits. Reducing the speed of computation will, of course, ease the heat dissipation problem, but it will also compromise the performance. Thus, computing solutions with reduced power consumption expressed per operation/unit of time are required. This might not necessarily be obtained solely with the technology but rather by using the technology within a computing architectural framework.

Form factors and weight are particularly important for mobile devices. A lot of research is ongoing to find possibilities to embed computing into items that people carry with them anyway – like clothes – resulting in wearable computing, or computing elements embedded in all kinds of everyday items. This is an area where the possibilities for using nanotechnologies to integrate functionalities into matter could be useful.

There is no doubt that the trend toward lighter devices and increasingly different form factors will be an important driver for future systems and this holds true for mobile computing. In this context, the technologies selected for the future should allow more freedom of design: even more compact and lighter systems, flexibility, moldability, and transparency will be major drivers.

Reliability will continue to be highly important for future personal mobile devices. Users are easily irritated by any failure in the devices that they use, especially if the failure prevents something important from happening. More reliable operation could be one driver for the introduction of nanotechnologies rather than the evolution of CMOS technologies: when reducing the line widths in CMOS to the nanoscale the reliability and thus the lifetime of the components will go down to levels that will become impractical for normal consumer devices. Perhaps other computing opportunities provided by nanotechnology can be made more tolerant to aging.

4.2   The digital mainstream

Today's standard computing system designs are based on two major pillars and span a wide variety of applications and usage, from mainframe to mobile computing. The first pillar is the CMOS semiconductor technology that supplanted bipolar technologies for digital design during the 1970s. The second pillar is the von Neumann, i.e., stored-program, computer architecture that is the basic structure for all modern information processing machines. Indeed, the meeting of CMOS and von Neumann architecture made possible the tremendous development of the computing industry.

Semiconductor information processing devices are among the most sophisticated, complex, high-performance structures. The first CMOS generation, the 10 μm generation, was created in 1971. Since then, and even earlier, the development of computing hardware has followed Moore's law for decades, with roughly a factor of 2 times smaller transistors every two years, leading to ever more effective and cheaper computing. This has led to significant improvements in economic productivity and overall quality of life through the proliferation of computers, communication, and other industrial and consumer electronics.

The most recent advancement is the Intel 45 nm transistor technology, which is already deep in the nanorealm. As line widths are reduced further the closer the fundamental physical limits will become. It is foreseeable that at some point CMOS will no longer be able to provide the improvements required to follow Moore's law, and new solutions will be needed. By 2020 silicon CMOS technologies are projected to reach a density of 10^{10} devices per cubic centimeter, a switching speed of 12 THz, a circuit speed of 61 GHz and a switching energy of 3×10^{-18} J. Is there really room for nanocomputing – can nanotechnologies somehow radically improve on these benchmark figures?

Due to the large investments in this field, the economical competitiveness of CMOS might last longer than its technological competitiveness. The roadmapping work of the current industry [24] envisages two major routes to following the Moore's law: scaling and functional diversification. Scaling includes a straightforward continuation in shrinking the physical size to improve density, decrease cost per function, and increase speed. Among the innovations that will enable this are: the use of extreme ultraviolet light (wavelengths shorter than 193 nm) in lithography; the use of metamaterials to produce the superlensing effect, allowing subwavelength lithography; and the use of new materials, such as high-k dielectrics to produce transistors with improved properties. Other types of improvements involve increasing the emphasis on local communications on chips, using of three-dimensional structures in addition to the traditional planar two-dimensional, parallel processing, multicore processors, and focusing on improvements via software. Even though transistor size is still following Moore's law, the benefits seen in realized products have fallen below previous expectations. One of the main reasons is that the high speed of Moore's law has led to less careful usage of memory and the number of machine microinstructions needed to carry out a basic function. The execution of a simple program that compares two objects in order to give an "identical" or "different" result today needs 50–500 times more system memory than was needed 25 years ago.

Functional diversification refers to adding functionalities that do not scale according to Moore's law, like nondigital functionalities migrating from system board level into a particular package-level or chip-level solution. Such nondigital functionalities include radio frequency (RF) communication, power control, passive components, sensors, and actuators – all of which are relevant for mobile devices.

But sooner or later the alternative computing solutions enabled by nanotechnology will be adopted. A natural step in that direction is to combine new nanotechnological elements on top of CMOS, with CMOS providing good interoperability with current hardware. In addition, the possibility of using operating principles other than charge-based logic needs to be investigated. Even if the evolution of CMOS and its combination with other charge-based computing elements can reach higher densities, the heat removal capacity will limit the obtainable performance. This is dictated by the fundamental physical limits.

4.2.1   The ultimate limits of scaling

In order to see what are the ultimate limits for computing systems one needs to derive the constraints set by the finite speed of propagation, information theory, the required operating temperature, heat dissipation, etc.

At nanoscales the devices become so small that the basic division between the computation and the physics by which it is realized can no longer be ignored. The efficiency of encoding digital information can absolutely not be reduced below the point where only one bit's worth of physical information – i.e., the available degrees of freedom present in the matter, determined by the number of atoms or elementary particles, and their properties – is being used to encode each piece of digital information. At the present rates of improvement, this will happen within the next few decades. Moreover, the physical size of devices operating with bits in normal atomic matter must also stop shrinking once a significant fraction of the physical information in the material is being used to encode individual digital bits. Once these limits start to become visible in devices in a plane, one needs to build devices in three dimensions. However, three-dimensional structures imply further challenges in heat dissipation, which is already at its limit in two dimensions, with the current technologies.

No energy, in the form of matter, information, or interaction, can propagate faster than the speed of light, approximately 3×10^8 m/s. Even though the speed of light is high, the limit it imposes on the computing speed will soon be reached: with a 100 GHz processing speed light can travel a maximum 1.5 mm distance and back during one cycle. In reality the speed of signal propagation will be less, and time is also needed for gathering information from memory. So in high-speed computing the architectures need to be more localized. This implies minimum communications and memory-access latency. Computers approaching this limit would require architectures featuring processing-in-memory. According to [12], meshes with 23(fractal) dimensions would be optimal, with the accurate dimension depending on the type of computing.

All commercial computers of today work in a nonreversible way. This means that all the energy required for computing, i.e., changing the physical state in the medium, is transformed into heat. Today this is already limiting the processing power, especially for small handheld devices for which the current limit is about 3 W. The required improvements include using physically reversible computing systems (adiabatic computing), computing with less energy, or using more efficient heat dissipation mechanisms.

The minimum energy needed for a binary irreversible switch is $kT\ln2$ and therefore the minimum amount of heat generated by one operation at room temperature is 3 zJ($zJ = J \times 10^{-21}$). On the other hand, to change the logic state a CMOS circuit requires transferring a certain amount of charge $Q = CV$, where C is the capacitance of the node and V is the operating voltage. This implies transferring an energy of $\frac{1}{2} CV^2$, which is of the order of 50 fJ with the current parameters ($V = 3.3$ V and $C = 10$ fF). Today this charge is returned to the ground terminal and the charging energy is converted to heat. If the computations were fully reversible the energy, including the thermodynamic and charging energy, could be collected back. It has been predicted that in the next 10 years reversible computing technology will begin to become more cost-effective than ordinary irreversible computing, and by the 2050s the difference will be of the order of 1000–100 000 [13].

The potential for fast switching and smaller switching energies is, in fact, one of the main advantages of new nanocomputing systems: in many other respects CMOS technology is already relatively close to the physical limits The operating temperature limits the maximum attainable rate of updating for traditional binary logic, i.e., the maximum clock speed or operating frequency. This limit stems from the fact that minimum energy required for one binary switch ($kT\ln2$) should be higher than energy uncertainty $\Delta E = hf/2$. Therefore, at room temperature, clock speeds greater than 8.7 THz are not possible [13]. Taking into account the limitations of heat dissipation, Nikolic *et al.* [32] have made calculations based on CMOS chip power dissipation: $P = NfpE$, where N is the number of devices, f the operational frequency, p the probability that a device is active during a clock cycle, and E the average energy dissipation per clock cycle per active device. If we keep 100 W/cm^2 as the limit for P, with $N = 10^{12}$ on a 1 cm^2 chip, and pE 50 times the room temperature thermodynamic limit kT, we get 2 GHz as the maximum operating frequency.

In conclusion, based on the physical constraints it can be shown that the optimum computational architecture for *charge-based classical digital computing* is a reversible, highly parallel three-dimensional mesh of processors [14]. However, all solutions for the implementation of computing systems are a compromise between several factors, such as cost, speed, energy consumption, size, and operating temperature. The right choice from among the available technologies depends on the particular needs, and the chosen physical implementation sets constraints on what kind of computing logic and architecture should be used. In the next section we look at more unconventional approaches to building nanocomputing systems.

4.3 New approaches for computing

Nanotechnology has created new tools for making things that could not have been made previously. In order to understand better what nanotechnology can offer for computing, it is essential to take a fresh look at what computing really is and how it can be done,

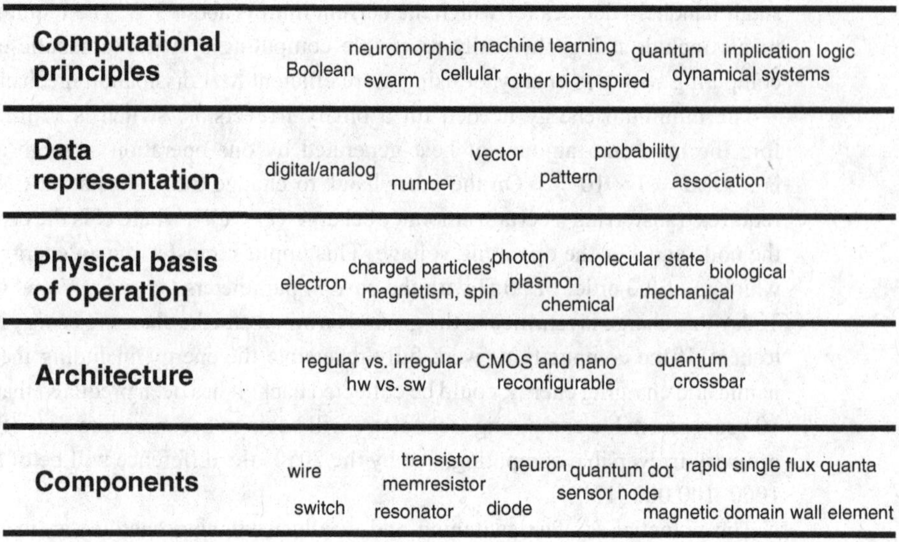

Computational principles	neuromorphic machine learning quantum implication logic Boolean swarm cellular other bio-inspired dynamical systems
Data representation	digital/analog vector probability number pattern association
Physical basis of operation	charged particles photon molecular state biological electron magnetism, spin plasmon mechanical chemical
Architecture	regular vs. irregular CMOS and nano quantum hw vs. sw reconfigurable crossbar
Components	wire transistor neuron quantum dot rapid single flux quanta memresistor sensor node switch resonator diode magnetic domain wall element

Figure 4.2 Hierarchical categorization and taxonomy for the new computing options.

starting from the basics of the theory of computation and the limits of physics, chemistry, and even biology.

Generally, "to compute something" means that we start from an initial state, perform an operationion on that state, and arrive at an output state, (usually) after a finite number of operations. To solve a computable problem one needs a way to store and read information, and to be able to operate on that information. In more detail, any algorithm can be solved by a Turing machine, the best-known mathematical model of computers, consisting of memory cells ("tape") to which information can be coded by a finite set of symbols, a "head" which can read or write information on the memory cells, and a table of rules according to which the current state and the input variable results in an output state and variable. In general, it does not matter how information is coded or what the rules are – any computable problem can be generated from these operations.

So a core question for nanocomputing is: are there physical implementations of computing states and symbols, computing rules, and computer architecture that would provide computing that is superior *in some respect* to the expected evolution of CMOS? This is not an easy question to answer. The maturity of the field is reflected by the fact that there have been only a few review articles published on the subject, with varying scopes and objectives. The ITRS 2007 review of emerging devices does discuss approaches other than CMOS or charge-based computing ones, but its main agenda is to compare them to general-purpose computing realized by CMOS evolutionary technologies. Here, we attempt to take a wider view, in particular keeping in mind the drivers and requirements of future mobile devices, as discussed in Section 4.1.

In order to select the most suitable solutions, one needs to understand the new options available. Understanding is easier, if the different options can be categorized and classified. Here we use the hierarchical classification presented in Figure 4.2, which has been prepared in the spirit of [4, 24]. This classification is explained below in more detail and the coming sections follow the hierarchical division described here. In these sections the focus is in nanotechnology, but general issues are also presented.

The topmost category is the "logic" used in the computation, meaning the computational principle. Present computers use Boolean logic, i.e., sets of elements consisting of the states 1 or 0 and the operations AND, NOT acting on the sets. Alternative options include, e.g., neuromorphic computing, machine learning systems, cellular nonlinear nets, and other dynamical systems, and even quantum; these will be discussed in more detail in the next section.

The selected logic gives boundary conditions on the next lower level, the data representation, which is discussed in more detail in Section 4.5. The first basic choice is between digital (discrete) and analog (continuous) variables. The current solution is digital, with analog systems also in use. But it might be advantageous to represent data in ways other than as just digital or analog numbers. Such representations could be vectors, patterns, probabilities, or associative connections between the data. These could be advantageous for efficient data size reduction to highlight essential features or to support alternative computing logics like neuromorphic pattern recognition. How this could be achieved in practice is an open question.

The logic and data representations have to be implemented in a physical form, which is discussed in the Section 4.6. Current computers are charge-based devices, more specifically based on the flow of electrons. Alternative physical bases for implementing the operation include magnetic states and interactions (spin, magnetic material), photons, plasmons, chemical interactions, states of atoms or molecules, mechanical or fluidical devices, and even biological material (neurons, viruses, etc.).

An integral part of any computing system is the governing architecture. This can best be described as the glue between the implementation technology and the application programming. Architecture can consist of identical elements forming a regular structure, or randomly organized elements. Architecture proposals for nanocomputing systems are reviewed in Section 4.7.

The lowest-level of the hierarchy consists of the needed components, such as switches, resonators, transistors, diodes, wires, interconnects between components, wireless sensor nodes, artificial neurons, quantum dots, magnetic domain wall elements, etc. This is the level on which most current nanotechnology research is concentrating. But, as mentioned before, the crucial question for efficient nanocomputing is not how to build superior components, but how to combine them together into circuits and systems which are optimized for the computation task or application.

4.4   Logical computation principles

As it is the current choice for computing, there is a lot of momentum and background work, supporting Boolean logic as the natural choice, especially for all-purpose logic. However, we should have a fresh look at alternatives, particularly for application-specific computation.

A powerful inspiration for many alternative computational approaches (both nanoscale and non-nanoscale) is the human brain. The brain is a very efficient computer in the sense that it functions in a massively parallel way and is very energy-efficient. The interconnect capabilities of brains are the key to their massive parallelism. The human brain has of the order of 10^{10} neurons and 4 orders of magnitude more connections between them, 10^{14} synapses.

One motivation for brain-inspired computation is to improve energy efficiency in artificial computers. Individual synapses in the human brain are relatively inefficient, consuming approximately 10^{-15} J of energy per bit of information that they generate; by comparison, in present-day silicon CMOS the switching energy is a decade less, only 10^{-16} J, and it is expected to have improved 100-fold by 2020 [23]. Overall, however, the human brain only consumes 10–30 W while carrying out functions that are impossible to emulate with present computers. Therefore, implementing brain-inspired computation could potentially lead to energy savings in man-made computers as well.

Another motivation is failure tolerance: in contrast to man-made electronics, living organisms are somewhat tolerant of even large failures, and use complicated but very efficient self-repair techniques. In artificial brain-inspired computing, failure of the components would not matter due to the parallelism and computational principles of neural networks; this is especially useful in technologies like nanotechnology where the failure rate is expected to be relatively high.

A formal model for a neuron consists of inputs associated with real-valued weights, modeling the synapses and dendrites, a device that computes the state of the neuron by summing the weighted inputs and applying a nonlinear function modeling the cell body (soma), and an output, which communicates the computed state of the neuron to the other neurons in the network that models the axon.

Neurons can be organized into networks with a wide range of topologies, ranging from networks arranged in feed forward layers, with forward propagation, for which a simple dynamic scheme is defined by an input $\rightarrow$ output relation, to the fully connected networks, called recurrent networks, whose many feedback loops endow them with great dynamic richness.

Two distinct process dynamics are applied to govern neural networks built in this way:

(1) *Relaxation dynamics* This calculates the state S_i of a neuron in terms of the states S_j of other neurons to which it is connected. This is expressed by

$$S_i = f(H_i);$$
$$\text{with } H_i = \Sigma S_j \cdot W_{ij} + \theta_j.$$

In the simplest case, the nonlinear function $f(x)$ is a Heaviside step function yielding binary states to the neurons. The variable θ_i represents a local weighting (threshold) at the neuron.

(2) *Learning dynamics* This calculates changes in interconnect weightings W_{ij}, where

$$W_{ij} = W_{ij} + D_{ij};$$

with $D_{ij} = f(S_i, S_j, H_i;$ other potential variable).

In contrast to states where the dynamics is well established, there are many algorithms for calculating D_{ij} from a wide range of parameters and error functions. The simplest learning algorithm is defined by Hebb's rule, where $D_{ij} = S_i S_j$ for signed binary states $-1, +1$.

However, crude it may seem, this model laid the foundations for a field that has developed considerably since the publication of work in 1981 by J. J. Hopfield [49]. By studying the dynamics of fully connected binary networks, called Hopfield networks, Hopfield showed that under certain conditions ($W_{ij} = W_{ji}$ and asynchronous dynamics), the network behaves as an excellent energy minimizer. This result complemented earlier work by Rosenblatt [50], who introduced the simplest neural model capable of elementary computation, namely, the perceptron.

The basic principles of neural networks can be generalized into a wider class of systems. The idea is to connect simple systems – which can be finite automata or, more generally, any dynamical systems – with a number of identical systems in a (local) manner, possibly with changeable types of interconnections. Interactions are usually local and the computational state variables can be discrete or continuous valued signals. When the number of identical systems grows large, the simple "cells" and their interactions can produce complex emergent behavior, and compute complex tasks. Cellular nonlinear networks and cellular automata are examples of these types of systems.

A commonly featured application area for cellular nonlinear networks presented in the literature is visual pattern recognition, i.e., pattern recognition in a two-dimensional array with input to each element in the array. As the end result is coded in the state of the whole array at the end of the calculation, and input is fed separately to each node as well, implementing the input and output in nanosystems will be challenging. Cellular automata – simple discrete versions of cellular networks, with nearest neighbor interactions – have been used as models of many physical systems (fluid dynamics, spin Ising models, biological systems, etc.), and in theoretical studies of computation (formal language recognition, for example), for generating random numbers, and in parallel solvers for certain types of arithmetic problems.

One class of these types of systems is message passing algorithms or belief propagation systems, which have been used in statistical physics to model collective phenomena like phase transitions, in information theory for error correction coding, and in general for optimization problems. There, the individual "cells" can hold different values (states) and exchange (probabilistic) messages with neighbors, which represent the constraints of the system. After iterations, the cells settle to a configuration that maximizes the probability/energy of the collective system.

Other biological systems besides the human brain have been the inspiration for novel ways of doing computations. Swarm intelligence is a term used to describe the behavior of specific types of networks of computational elements. Swarm computing systems consist of a multitude of simple, decentralized elements with limited functionality, connected with each other (locally) to produce complex emergent behavior. Such computing algorithms were originally inspired by biological systems like ant colonies and the collective behavior of a flock of birds. The algorithms usually update their state based on local conditions, velocity (gradient of the local state), and some type of memory of previous states, to optimize a global reward/quality function. Application areas under research include optimization [11], data mining [1], and classification.

An evolutionary or genetic algorithm is a method of adapting to the environment by mimicking natural selection. Genetic algorithms use combining and mutation of their parameters or even algorithmic elements, and test the resulting algorithms against predetermined quality criteria. They can be very efficient in certain optimization tasks where other algorithms might not perform as effectively. As an example, in [72] genetic algorithms are used to optimize the cost of circuits.

A more general framework for adapting and learning algorithms is machine learning [73], a field of research in which models are automatically learned from sets of example data (observations), instead of an expert constructing the models manually beforehand, as with many of the above–mentioned methods. Some machine learning algorithms are partly robust to incorrect values of input parameters, and are thus suitable for use in systems where inaccuracies are inherent in the systems they are simulating or the systems on which they are implemented. Machine learning methods are very efficient in pattern recognition tasks, or any other classification problem where the data need to be sorted into discrete sets based on often incomplete information. An example in which the possibility of nanocomputing to implement machine learning to ascertain whether radio channels are free or occupied can be found in [75].

The principles of quantum computing are determined by the behavior of quantum systems. The superposition and entanglement of quantum states allow fundamentally different ways of representing data and performing operations on them. In a quantum computer, the data are measured as qubits, i.e., states of a quantum system, such as spin up or spin down states of an electron. Quantum computers can be modeled with the quantum Turing machine, i.e., the quantum version of the famous Turing machine described above.

Although quantum computation holds great promise for certain applications, there is no reason at present to assume it would perform better in general computation tasks than traditional computers. The most important applications in which quantum computing could offer significant advantages over digital computers are factoring problems, certain optimization problems, calculations of a discrete logarithm, and in particular simulations of quantum physics systems. Of these the first one is the most obvious reason for intensive work on quantum computing: the current encryption algorithms are based on the fact that factoring large primes is very slow with digital computers, whereas with quantum computers factoring could be performed significantly faster.

4.5   Data representation: analog or digital?

For quite some time a major trend in electronics has been towards digitalization. However, as discussed in the previous section, computation can be performed in many ways, and with different computation principles digital data presentation may not be the best choice.

Nanotechnology could increase the importance of analog processing due to at least two reasons: the first reason derives from the new possibilities of measuring the real analog world. As the sensor outputs are in analog form, it could be very inefficient to transform this huge amount of information directly into digital form. Analog-to-digital converters (ADCs) are very power consuming. ADCs have failed to keep up with Moore's law: their performance has doubled only every 8 years instead of the $1\frac{1}{2}$ years of processors. If no breakthroughs happen for ADCs, the interfacing with the physical world could become a major bottleneck in the power consumption and cost of digital systems. A more efficient solution could be to integrate analog processing with the sensors to compress the information for potential transfer to digital form.

A second reason for the potentially increased importance of analog processing is simply the fact that with the new opportunities provided by nanotechnologies the rules of the game have changed. The cost/benefit considerations which have been valid until now may soon be reversed, and there may be new physical realizations based on nanotechnologies that will help to overcome the difficulties of analog processing in a natural way.

4.5.1   Differences between analog and digital processing

The main driver of the digital domination has been robustness against noise. The accuracy of digital calculation is determined by the number of bits used to represent the numbers. The noise from round-off errors can always be diminished by increasing the number of bits used for representing real numbers. But since representing one bit in a digital system requires one (discrete) physical state the increased accuracy comes at the cost of silicon area, speed, and more memory required. In contrast, with analog systems continuous states are used to represent the numbers. Therefore the accuracy is determined by the quality of the components – which is always a challenging issue in any mass manufactured system.

Analog circuits generally use less power per operation than corresponding (electrical) digital circuits. This is partly due to the robustness of digital systems: the digital quantities are represented by voltage or current values which need to have enough separation to make them robust – this requires large enough current or voltages. Also, an analog device uses only one voltage value for each number, instead of the several values needed to represent a digital number with the required accuracy. An analog circuit only uses power for fast switching when the signal it is processing changes quickly. Therefore the operating frequency for an analog circuit can be lower than the frequency required for digital processing of the same waveform. As a result, analog circuits tend to use

10–100 times less power, and several times higher frequency waveforms than their digital equivalents [40].

One of the main differences between digital and analog systems is that digital processing is usually done sequentially or serially, whereas analog computing is simultaneous or inherently parallel, without needing to resort to multiple processors. Digital computation usually requires programmed instructions, which need to be stored in a memory unit – this can prove to be a major bottleneck in the processing, since it allows only one arithmetic operation to be performed at a time. As a result, an increase in problem complexity, i.e., an increase in the number of arithmetic operations, requires an increase of the processing time. For analog processing the time is independent of the problem complexity, and depends only on the physical realization of the computing unit, i.e., characteristics of the electronic components and input/output devices.

Therefore, the whole system architecture for an analog computer is determined by the types and numbers of operations needed, making them application-specific. In contrast, digital computers tend to be general-purpose machines, with changeable programming. This is a major drawback of an analog system: even if analog systems can perform general computations like von Neumann computers do, the constructions required are often quite cumbersome and inefficient. But for applications like sensory data processing and radio signal processing where speed or the complexity of the problems is the limiting factor, application-specific circuits can be a competitive choice.

Another problem with analog systems is that they are harder to design and debug due to the lack of mature automated design tools. Also, in the analog domain, memory is more challenging to implement: analog quantities are harder to store reliably. However, for certain computing principles and physical realizations, the concept of a separate memory is not significant: e.g., neuromorphic systems can store information in the same elements that are used for computing (the "strength" of connections to other neurons).

4.5.2   Computations with dynamical systems and hybrid analog–digital systems

The choice between analog and digital computing depends on the application, chosen computation logic, and the physical realization implementing it. The optimal solution would combine the robustness and easy programming of digital computing with the power efficiency and speed of analog methods. In fact, theoretical comparisons of analog and digital computing suggest that the most efficient use of computing resources is achieved with a hybrid approach, combining analog and digital in the same computation [35]. One very promising possibility is to use fast analog processing elements with digital feedback or error correction for increased accuracy and robustness.

In addition, the general concept of dynamical systems offers rich possibilities for computing, which have not yet been fully explored. By constructing a suitable dynamical system and implementing it in an analog device, one can perform quite intricate computations, with the robustness usually associated with digital (non-continuous signal) systems, and one can also perform general computation, i.e., realize general automata and universal Turing machines [8, 5]. Brockett has constructed dynamical systems which take as input continuous signals/pulses, but produce as output discrete values, depending on the generic homotopy class of the input signals. There are also other interesting examples of dynamical systems with robust output properties, which can, e.g., sort lists [7], and act as shift registers [6] or as flip-flops [17].

These types of systems can therefore be used as design primitives for basic computational elements: the robustness is due to the discrete output, but unlike digital systems that suffer from the finite presentations of real numbers, these types of systems could produce a more direct method of interfacing the analog and digital worlds.

Systems of nonlinearly coupled oscillators are other interesting examples in which computing with dynamical systems is performed in a robust manner. Simple systems of nonlinearly coupled oscillators can produce very accurate frequency synchronization, regardless of the initial state of the individual oscillator elements [26]. Whether the synchronization will occur depends on the input values, i.e., the couplings between the oscillators. This sort of system can act as an efficient detector as the collectively synchronized mode produces a robust signal which is easy to detect.

These types of systems can be viewed as alternative examples of how hybrid digital–analog systems could be realized, analyzed, and designed in a more optimal manner. There exists a rich theory of control of hybrid systems, based on the generalized theory of dynamical systems, which is outside the scope of this survey. However, the ideas discussed in this section should be kept in mind when looking for new ways of processing information with unreliable nanoscale elements, as the resulting systems are likely to incorporate both digital and analog processing.

4.6   Physical implementation

The chosen logic and the data representations have to be implemented in a physical form. Obviously, neither all data representations nor all kinds of logic can be implemented in all physical systems and hence often the possible physical basis sets the limits for the other hierarchies. However, in general, the fundamental choice when designing any physical implementation of computing is to choose the medium (i.e., states), and how to realize the basic interactions (computational operations). These set the limits for the overall performance of the system.

Although nanotechnology has the capability to improve physical implementations such as CMOS, the main question that remains is what can new nanotechnology bring to the computational field? One of the greatest promises is that it will enable operations with individual physical quanta, e.g., single electrons, single spins, molecular states, flux quantas, or even with quantized mechanical states of nanoelectromechanical resonators (NEMS). This could yield, e.g., the ultimate information density of computing devices, as calculated in Section 2.1. Nanotechnology also provides physical systems with an extremely small interaction energy, i.e., the energy required for computing operations. The usual requirement for the physical system is, however, that the energy required for computational operations should exceed the thermal energy. Yet subthreshold solutions, such as stochastic resonance, do exist. Stochastic resonance is described in more detail in Chapter 5.

In highly integrated applications it would be essential to use the same physical bases for the computing as the system uses elsewhere for other purposes. For instance, in some sensing applications it would beneficial if the transducing, the data gathering, and the primary information processing could be done using a similar or the same physical structure. On the other hand, the physical implementation of the interconnections and the general architecture limit the possible computing bases as well.

In addition, the possible choices of manufacturing techniques set constraints on the physical basis of operation which can be used. For instance, a regular computing architecture would most naturally be realized by using a bottom-up approach, i.e., making computing structures by self-assembly. This limits the choice of possible materials to those allowing self-assembly technique to be used.

Nanotechnology is an extensively studied field of research and there are numerous physical systems enabling computing structures. The complete list of all the possibilities is beyond the scope of this book, but the choices for physical systems that can be used for computing operations include:

(1) Charge-based devices, i.e., devices which operate with manipulating electric fields (electric potential, charge flux). All modern computer implementations, aside from some memory components, are based on electronic devices.
(2) Devices based on atomic or molecular states, like controlled excitation and de-excitation of atoms, molecules, or molecular conformation.
(3) Devices based on magnetic interactions, like spin or more macroscopic magnetic fields. Magnetic devices can be constructed so that they do not require any power when not in active operation, without losing their current information state. Being ready to operate without booting up would be a major advantage in any computer system.
(4) Devices based on flux of photons, or their polarizations.
(5) Devices based on collective excitations of, e.g., photons and electrons, i.e., plasmons.
(6) Devices based on chemical interactions. The problem with these is that chemical reactions are very slow when compared to the speed of electric and magnetic inter-actions between simple particles like free electrons. An advantage could be that it is very easy to produce massive amounts of the elementary "devices," i.e., chemical compounds/molecules, such as DNA, to realize massively parallel processing.
(7) Mechanical or fluidical devices which are operated by forces producing spatial displacements or phonon excitations. The first real computers were actually mechanical devices designed by Charles Babbage as early as the beginning of nineteenth century. Also computing devices based on moving liquids are possible.
(8) Biological devices based on biologically produced entities, like neural cells or viruses.

In Figure 4.3 examples of different physical implementations are illustrated.

Electronics and other charge-based devices are the current choice with a lot of momentum. The semiconductor community claims that it does not make much sense to invest heavily in conventional charge-based logic as the new solution for computation

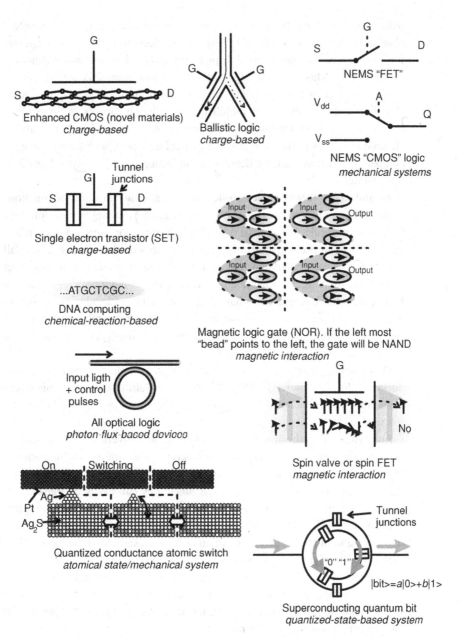

G

S ━━━━━━━━━━━━━━━━━━━━━━ D
Enhanced CMOS (novel materials)
charge-based

Ballistic logic
charge-based

NEMS "FET"

NEMS "CMOS" logic
mechanical systems

Tunnel
junctions
G
S   D

Single electron transistor (SET)
charge-based

...ATGCTCGC...

DNA computing
chemical-reaction-based

Magnetic logic gate (NOR). If the left most
"bead" points to the left, the gate will be NAND
magnetic interaction

Input ligth
+ control
pulses

All optical logic
photon flux based devices

Spin valve or spin FET
magnetic interaction

On   Switching   Off

Ag
Pt
Ag_2S

Quantized conductance atomic switch
atomical state/mechanical system

Tunnel
junctions

"0" "1"

$|bit> = a|0> + b|1>$

Superconducting quantum bit
quantized-state-based system

Figure 4.3 Some representative examples of the possible physical implementations (clockwise from top left): enhanced CMOS solutions based on novel materials, Y-branch ballistic logic (e.g., [34]), logic based on nanoelectromechanical switches (e.g., carbon nanotube [15, 16]), magnetic logic gate for binary computing [19], spin valve transistor [31, 42], superconducting quantum bit based on phase across tunnel junctions [41], quantized conductance atomic switch [38], all optical switch based on a silicon microring resonator [43], computing based on DNA [36], single electron transistor based logic. S stands for source, G for gate, D for drain, V_{dd} for drain voltage, and V_{ss} for source voltage.

with nanotechnology, because, e.g., the heat removal capacity will ultimately limit the obtainable performance [23, 44]. One conclusion from this is that one should look for alternatives to charge-based approaches rather than try to compete with CMOS in this arena. This forecast does not include new materials for transistors, such as CNTs and graphene, which could surpass silicon for low-power applications requiring high speed, due to the larger mobility of the charge carriers in these new materials. Furthermore, nanotechnology, e.g., graphene or CNT technology, could also be used for electronic devices based on ballistic electrical transport, with competitive prices when compared to current ballistic devices (as an example see [34]) based on GaAs/GaAlAs heterostructures.

In molecular (atomic) electronics individual molecules are the functional components. For example, transistors can be constructed from single molecule. Even though the active molecule may consist of only a few atoms (or even just one), it is difficult to build small devices using molecular electronics because they require such small contacts. Currently the poor manufacturing control and reliability mean it has only been possible to produce individual components for testing in the laboratory by chance after many failed attempts. Using a cross-bar architecture with molecules at the intersections of the grid might allow a controlled arrangement, operation, and manufacturing [28]. Also bottom-up methods and reconfigurable architectures may provide successful approaches for molecular electronics in the future.

One interesting approach for molecular or atomic electronics is the quantized conductance atomic switch (QCAS). QCASs act as atomic-scale mechanical relays, where the atomic-scale contact formation and annihilation are by a controlled electrochemical reaction [38].

Spintronics uses the orientation of individual spins to encode information in electrons or nuclei for storage, logic, and communications. One option is to orientate the spins of particles then transfer them to a reader to detect the orientation. Usually particles with charges, most often electrons, are used, since they can easily be transferred from one location to another by electric fields. Challenges include orientating the spins of a large enough percentage of the particles, then keeping the spin orientation coherent during transport, and finally detecting the orientation. An example of the basic building blocks of spin logic is the so-called spin valve ("spin-FET") [31, 42].

Another option is to induce spin waves, collective oscillations of spins around the direction of magnetization [25]. Such spin waves can also transfer information over small gaps and enable wireless interconnects. Due to exponential decay of the signal, the gap is in the submillimeter range.

The potential advantages of spintronics include low-power electronics with a single chip including all functions of storage, logic, and communications, CMOS compatible fabrication, and spin wave error tolerance to imperfections with characteristic sizes less than the spin wave coherence length. Challenges to be solved include efficient injection of spin-polarized electrons into nonmagnetic materials, controlling the loss of spin polarization or spin coherence, i.e., fast decay of signal, and efficient spin detection.

Magnetic logic based on nanoscale magnetic objects could offer a more robust method than the manipulation of individual spins. Nanoscale magnets are not subject to stray electrical charges, and they are easier to fabricate and manipulate than individual spins. Magnetic storage devices have been around for a long time, and information processing based on patterned magnetic media has been proven possible. For example, **quantum cellular automata (QCA)** devices have been realized using nanomagnets [9]. Other physical realizations of QCA have been constructed, e.g., from molecules [27], nanomagnets, and quantum dots. The interactions between the cells are then based on chemical reactions, magnetic forces, and entangled quantum states, respectively. A common feature of all these systems is an organized grid of cells.

An interesting example of magnetic information processing has been realized by researchers at Notre Dame University. Their concept is based on the electrostatic repulsion of charges in an elementary cell made from four magnetic quantum dots [21]. Connecting such cells and exploiting their interactions, one of the two stable states can be propagated along a circuit set up rather like a series of dominos. A whole range of binary logic 'circuits' (AND, OR, XOR, INV) has been described on the basis of this idea (see, e.g., [19] and references therein).

Because the chip has no wires, its device density and processing power may eventually be much higher than transistor-based devices. Moreover, the energy consumption is much smaller, reducing the need for thermal management. Devices using these types of magnetic chips would boot up almost instantly, since the memory is nonvolatile, i.e., it retains its data when the device is switched off. However, the operation is not very efficient or fast when compared with more traditional ways of realizing Boolean logic.

Plasmonics uses surface plasmon polaritons to manipulate information. Plasmons are collective oscillations in electron density. The collective oscillations of surface charges (surface plasmons) can strongly couple with electomagnetic waves in the optical regime leading to the excitation of surface plasmon polaritons.

Plasmons have been considered as a means of transmitting information on computer chips much faster than present techniques can, since plasmons can support frequencies into the 100 THz region, while conventional wires become very lossy in the tens of gigahertz region. They have also been proposed as a means of high-resolution lithography and microscopy due to their extremely small wavelengths. Both of these applications have been successfully demonstrated in the laboratory.

Surface plasmons have also the unique capability of confining light to very small dimensions which could enable efficient integration of optical and electronic components. One bottleneck with current computing chips arises from the interconnects and surface plasmons could provide an efficient means for such wiring [33].

The interesting high-frequency properties of plasma waves in two-dimensional electron gas (2DEG) systems have been studied extensively for semiconductors. The 2DEG can serve as a resonant cavity for the plasma excitations, or act as a voltage controlled waveguide which can be used in terahertz devices such as radiation sources, resonant detectors, broadband detectors, and frequency multipliers [10].

Molecular or chemical computation is based on sets of chemical reactions. For instance, sets of chemical reactions can be used to construct finite automatons, where chemical compounds correspond to states, and operations (reactions) can be triggered by external stimuli like light of a certain wavelength, the chemical concentration, or the temperature.

The main advantage of chemical or molecular computation is the possibility for massively parallel processing with very low power. This is useful especially if the computation time grows exponentially with problem size, or when the power requirements are very restrictive. Challenges with this approach are automatization of the computation, poor error tolerance, difficult realization of input and output methods, slow speed determined by the speed of chemical reactions, and most importantly, the fact that a practical mathematical problem that would justify the use of massive parallelism has not yet been identified.

Nano(electro)mechanical computers have also been studied extensively. In one concept, mechanical digital signals are represented by displacements of solid rods [30] and the speed of signal propagation is limited to the speed of the sound, e.g., 1.7E4 m/s in diamond. This kind of device could have very low power requirements due to the extremely small masses of the nanoscale elements: about 1 zJ per operation. This would be five orders of magnitude less than with silicon CMOS in 2005 and three orders of magnitude less than what is estimated for silicon CMOS in 2020. The reason for this is that mechanical computers can more easily be made energetically closely reversible, e.g., because spring forces can store the energy needed to move the rods. What is interesting about these numbers is that 1 zJ per operation is less than the thermodynamic limit of $kT\ln 2 = 3$ zJ for irreversible computation.

One limitation of this kind of approach is that it gives clock speeds only in the kilohertz–megahertz range [23]. It is also expected that manufacturing challenges for nanoscale moving parts will be significantly higher than for stationary nonmechanical systems. There might well be applications for this type of ultralow power computing, which can be very tolerant to a wide range of temperatures and conditions, e.g., in sensor applications.

A related approach is to use NEMSs for computing. For example, coupled nonlinear oscillators can be used for various signal processing tasks, and with NEMSs it may be possible to reach the desired frequency ranges [71]. On the other hand, a network of coupled NEMS (or microelectromechanical MEMS) resonators has been proven to be able to act as a neural computer with pattern recognition capabilities.

One interesting approach is to fabricate a logic which acts as a conventional CMOS logic, but is using mechanical switches (e.g., CNTs [15, 16]) instead of solid-state semiconducting field effect transistors (FETs). The main advantage of mechanical switches is that power consumption is virtually zero in the absence of operations. On the other hand, the integration of the switch logic with the present CMOS logic is straightforward if the computation principles are similar.

DNA computing is a good candidate for chemical molecular computation since the ordering of the bases in DNA can be used to code and manipulate information. In particular, the importance of DNA in other fields of science has created technologies to manipulate DNA relatively easily, also in an automated manner. In December 2006 modularity and scalability of DNA computing were demonstrated when eleven DNA gates were combined into a larger circuit [36]. Logic values 0 and 1 were represented by low and high concentrations of short oligonucleotides. Because input and output were represented in this same form, the construction of multilayer circuits was straightforward.

The power requirement for DNA computing is estimated to be 1E-19 J per operation, which is three orders of magnitude less than that for CMOS [2]. But due to the high cost of materials and the slowness of diffusive molecular interactions, DNA computing might never be a cost-effective computing paradigm [12], except for special-purpose computing tasks performed in chemical or biological media.

4.6.1 Recipes for nanocomputing systems

The list of examples presented above is by no means exhaustive – there is certainly room for new implementations of computing systems. For a nanotechnology researcher, the natural starting point when searching for new realizations of computing devices is to look at the physical phenomena and to base on those. Here we attempt to present some ideas on ways of designing novel types of systems that can perform computation in some manner.

To start with the simplest examples, the first thing one can do is to look for a system with a discrete set of states which are stable and robust enough to use, and for some mechanism to include transitions between states using external stimuli. If the states can be connected to form a finite automaton the system can already act as a simple computer. The external stimuli can be electrical, magnetic, mechanical, light or other electromagnetic radiation, chemical, or even biological, etc. The transitions have to be reliable, repeatable, and fast enough for the targeted application. This is by no means an easy thing to realize, as the requirements for most practical applications are quite severe, particularly for mobile systems. Small size, low energy consumption, and ease of input and output of are always features to seek.

More complex computer systems are obtained by connecting together simple systems with elementary computing action, i.e., state change action, giving a network of simple elements. In this way, simple "circuits" can be constructed. Or, if the interactions between the elements can be tuned, large systems can exhibit collective emergent behavior. These types of systems could be used, e.g., for implementing belief propagation algorithms. Examples of analog electronics implementation of message passing algorithms for wireless communication signal processing (forward error decoding algorithms and Kalman filters) can be found in [40]. Other examples are the magnetic cellular automata, in which the states are the orientations of magnetic particles and the interactions are the magnetic interactions between them.

To generalize this to continuous systems one can look for interesting dynamical systems, i.e., connected (nonlinear) partial differential equations governing the behavior of the system. If the interactions and nonlinearities can be tuned by external means, you already have a system with potentially very rich behavior. Does the system have any attractors, stable minima, chaotic behavior, etc. at parameter values? A system close to a bifurcation point can be, with very small changes in input, driven to different states, and thus used for classification and triggering systems. The search for a suitable physical realization has to start from the equations governing its behavior, or states and transitions of the computational problem.

Another route is to think of what the basic components for realizing a particular computation principle are. For Boolean logic, basic logical operators AND and OR are needed. These operations are done with controllable switches, e.g., transistors. However, a switch can be realized in many ways, not only with the electrical field effect in semiconductors. For signal processing in general, one can try to realize basic functions like sorting values, shift registers, delay lines, or in fact any of the functions previously done with analog electronics or optics. Good oscillators are always needed, and efficient methods for performing analog-to-digital conversion (or vice versa) are very valuable. A particularly useful feature in a nanosystem would be the ability to perform multiplications in a simple manner: addition can usually be realized easily (in electrical circuits connecting the wires), but multiplication often requires active elements (like transistors). With addition and multiplication one can approximate nonlinear functions by their Taylor expansion – enough tools for solving many interesting problems.

However, the most difficult questions in building efficient nanocomputing systems are related to the system architecture, not to individual components. Not only does the architecture have to be efficient in realizing the states and connections between them, but it also has to be such that it can be manufactured easily. Without these characteristics no computing system can successfully compete with current technologies.

4.7 Architectures for nanosystems

The crucial test for all computation approaches comes at the system level. The end result is a combination of the chosen computational principle, data representation, and their physical implementation, down to the level of individual components. What ties these together is the architecture of the computing machine, or its actual organization of circuits and/or subsystems. According to [32], the incorporation of devices into larger circuits reduces the computational speed from device level to 100 or 1000 less at the system level. So in order to get any benefit from nanocomputing systems, the architectures have to be considered and designed very carefully.

The archetypal computer with which we are so familiar today is based upon a principle stated by Alan Turing in 1935, when he first described the famous machine which now carries his name [46]. This purely conceptual invention was developed at the time to illustrate his mathematical research in the field of logic, computability, and other concepts in computation theory. The idea was taken up and adapted a decade later by John von Neumann and the result gave birth to what is now the basis of the computer structure as we know it [47]. This type of computer structure is known as the "von Neumann architecture." It is the common "reference model" for specifying sequential digital architectures, in contrast with parallel architectures. In the von Neumann architecture, a processing unit is connected to a memory device which stores the data to be processed and a step by step list of all processing instructions. The sequence of processing steps constitutes the application code to be executed. A control unit synchronizes the read and write operations between the processing and the memory unit. To be complete, the architecture includes a device that handles communications with the outside world.

The von Neumann architecture is characterized by the fact that the processing steps are sequentially ordered (one step after another), and that data and program instructions are stored together in the same memory. This actually constituted the key refinement made to the Turing machine by von Neumann. Not all instructions that are executed actually process data. For example, some control instructions are used to manipulate the sequence itself, effecting operations such as omission, branching, and stopping. These essential instructions are used to write more complex application algorithms than a simple linear series of instructions.

Memories, interconnects, and operators are thus the basic elements for building a computer according to the model we have just described. We may note in passing that, of these three functions, only one – the operator – can modify the state of the data. Memories and connections must in no way alter the state of the data they handle.

All in all, the von Neumann structure made of CMOS implemented logic gates, memories, and metal interconnections and fitted with the algorithmic programming model acting on Boolean coded data made a rather great recipe. All the elements fit perfectly together.

However, the conventional architectures with regular structures are not necessarily the best suited for the nanoscale world. A computer architecture organized into well-defined blocks leads inevitably to lengthy nonlocal interconnects and implies a structure with a high level of granularity. This is particularly relevant for the control unit and operator elements which possess a rather irregular internal structure. This irregularity in the distribution of interconnect circuits complicates the timing control of signals. Also, a computer made in this way is very vulnerable to defects. Whether one is dealing with fabrication defects, breakdown, or dynamical perturbation due to noise, regular architectures are not robust in the face of mishap. In fact, perhaps the most important reason why nanocomputing systems have not been realized in practice yet is the lack of a good architecture, with easy manufacturing and robustness against defects and failures.

4.7.1 Challenges of nanoarchitectures

The von Neumann paradigm differentiates between memory and processing, and therefore relies on communication over buses and other long-distance mechanisms. This throughput bottleneck due to (single) memory access has become one of the limiting factors for performance improvement. As a result it has been predicted that the focus of microarchitectures will move from processing to communication [3]. However, nanoelectronics could fundamentally change the ground rules, as with it processing could be cheap and plentiful, but interconnections expensive. This would move computer architectures in the direction of locally connected, reconfigurable hardware meshes that merge processing and memory.

With CMOS the manufacturing failure rate per device is about 10^{-7}–10^{-6} [32]. All nanodevices are to a greater or lesser extent hard to make reliably with failure rates dramatically larger than for current CMOS. Failure rates have to be and will be improved, but might not reach anything like that for CMOS. High failure rates seem to be a fundamental phenomenon in nanoelectronics that one needs to live with. Different strategies for coping with this are:

- Using redundant computing elements, e.g., triple modular redundancy, in which the same operation is performed by three devices and the majority opinion on the end result wins.
- Testing the operation of all devices on a chip and reconfiguring the circuit after manufacturing. An example of this is the Teramac computer, in which a simulation is done with conventional field programmable gate arrays, FPGAs [45].
- Use of algorithms that are tolerant to errors in the computing hardware, like different forms of parallel computing, e.g., cellular automata or machine learning algorithms.

The first option is more in line with the current 'hardware approach' to computing: the aim is to create the wanted hardware with high reliability. In this case the manufacturing is hard but the programming is more straightforward. With decreasing size, inventive manufacturing methods need to be developed in addition to the current lithographic techniques. Here, for systems with completely regular layouts, self-organization or self-assembly could be used. As an example, many approaches with cellular automata and cellular nonlinear networks can be completely regular. For more irregular designs, methods like DNA origami techniques could be the solution.

Due to the difficulty of high reliability manufacturing of large quantities of nanoscale components, it makes sense to consider also randomized or partly randomized manufacturing techniques. Even though the manufacturing methodology would not be based on any randomness, there are likely to be more errors the smaller the targeted components are. The emphasis then will be more on the "software approach," i.e., reconfiguration and programming afterwards to get the desired functionality, or on algorithms which are robust against irregularities and failures of hardware. In contrast to man-made electronics, living organisms use complicated but very efficient self-repair techniques. This would of course be the ideal solution to deal with errors and failures, and it can to some extent be replicated by algorithms that can constantly learn, reconfigure, and adapt to the changed situation. Interesting possibilities here are machine learning methods. Given sufficient knowledge about how a nanoelectronic circuit is intended to be used, machine learning methods could potentially be used to find a design for the circuit that is more robust to failures of components and connections than existing human-invented circuit designs. On the other hand, due to the robustness of machine learning methods against parameter inaccuracies, nanoelectronic circuits that implement these types of methods could still be useful despite physical failures, more so than circuits implementing less robust methods. Thus, machine learning methods could be useful applications for fast nanoelectronic circuits. The usefulness of machine learning in nanocomputing has been reviewed in [39].

Besides failure tolerance and manufacturing problems, a difficulty in nanocomputing systems is finding efficient input/output methods. This will be particularly challenging for molecular-scale electronics. In this one of the few realistic input/output solutions would be based on light: using fluorescent or photosensitive molecules it is possible to

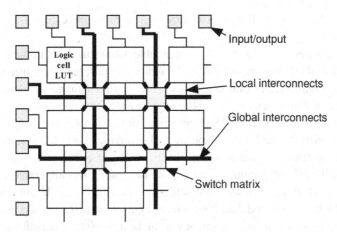

Figure 4.4 The typical structure of a reconfigurable architecture.

interface directly between nano- and macroscales, and even to integrate with conventional electronics. For some time to come, nanocomputing systems will need to interface with conventional electronics, so it is crucial to realize the interfaces so that they will not become a major bottleneck in the system.

Also, it is likely that new approaches will be needed for efficient interconnects *within* the system: e.g., at the nanoscale metal wires become very lossy, and can easily degrade the efficiency of the system. Suggested solutions for these problems include other nanostructures in wiretype form (e.g. nanotubes), Coulomb-coupling, magnetic coupling, and electromagnetic radiation between "nano-antennas."

To date, there are no suggestions for nanoarchitectures which would successfully solve all these problems. In the following, we review ideas on how to address these issues.

4.7.2 Examples

Architecture for reconfigurable computer systems is based on the idea of implementing Boolean computational functions in a look-up table (LUT) structure. A LUT is just a memory bank in which the content has been programmed according to function $f()$ in such a way that $D_o = f(A_i)$, where D_o (the output) is the memory data bus and A_i (the input) is the memory address bus. As well as the LUT structure that implements the actual functions, a reconfigurable computer also includes a programmable interconnection network (switch matrices). The combination of simple operators implanted in the LUT units via the communication network can then be used to configure any type of rather complex logic circuits. Adding input/output facilities to this scheme, one obtains a typical structure of a reconfigurable circuit like the one shown in Figure 4.4.

A set of memory cells is associated with each resource in the reconfigurable architecture: the LUT for the logic blocks, and a specific memory cell to drive the state of each switch in the interconnection network. The whole system (logic plus interconnects) is therefore fully programmable by virtue of writing those memory cells. This type of architecture is the basis of all current commercial reconfigurable circuits, commonly called FPGAs.

One of the main advantages of the reconfigurable architecture lies in the fact that it is essentially based on *memory technology*. In addition to this, the array structure of the logic operators within an interconnect network yields rather regular geometries. For these reasons, the architecture of a reconfigurable computer has been proposed by many experts as a natural channel for nanotechnology implementation. This idea was first popularized through the Hewlett-Packard project known as the Teramac [48]. The original aim of the Teramac project was to take advantage of the notion of circuit reconfigurability to enhance the reliability of computers. Indeed, by integrating reconfigurable circuits at the hardware level, it was foreseen that self-repair facilities could be integrated into computers. The results showed that, if such reconfigurable structures could increase the intrinsic reliability of a system, they might well make it possible to exploit components produced using technologies that were a priori rather unreliable, such as ultimate CMOS or nanotechnologies.

As reliable manufacturing is a challenge in nanotechnology, one could also build hardware that is possible to reconfigure to perform the required operations. Use of reconfigurable logic requires the use of more surface area on a two-dimensional chip. An estimate with silicon CMOS is that FPGAs result in a 20–50-fold area penalty and 15-fold delay penalty when compared to heterogeneous, custom-designed solutions. Of course, nanotechnology will always require, at least with the present understanding, more space on the chip than the theoretical minimum solution, due to manufacturing and operational errors.

Using reconfigurable logic in nanoelectronics is aimed at a regular hardware structure that is easy to manufacture. The complexity is then moved into the reconfiguration and operation by software. A widely studied hardware here is the array architecture with crossing nanowires: since orthogonally overlapping wires reduce the need to precisely align the wire it will be the choice of design largely used at least until more advanced manufacturing methods arrive. Two-terminal devices created by these types of cross-bar structure are easier to manufacture than traditional transistors, as it is easier to bring two terminals into proximity than three. However, three-terminal devices could bring more functionality. For example, having two-terminal devices in the cross-bar intersections results in the inability to achieve signal gain, at least by the methods considered in [37]. This limiting of gain would restrict the array size. Due to different scales in nanosystems and many traditional approaches, the connections between these two parts need to be sparse.

In most cases in the near future, combination of CMOS and nanosystems could be a basic requirement due to the need to connect the nanodevices to other forms of electronics or to manufacture them. One approach would be to add nanoelectronic components to the CMOS structure. Another approach would be to have the nanoelectronic structure as the main computational platform and use CMOS to make connections or to allow easier fabrication.

An example of the latter case is the CMOL architecture proposed by Likharev and his colleagues [28, 29] (Figure 4.5). CMOL combines CMOS devices with crossed

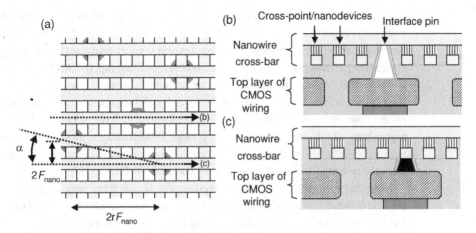

Figure 4.5 The CMOL approach. Reproduced with permission from [29].

nanowires and molecular devices. Fabrication is done in two layers, with one layer consisting of CMOS blocks or cells and the other layer containing an array of crossed nanowires as interconnects between the CMOS cells. The nanowire cross-points contain programmable molecular devices, permitting reconfiguration of the nanowire-based connections between the CMOS cells. Two types of applications for CMOL circuits have been proposed: neural networks and FPGAs. However, no fabrication experiments have yet been done to build and test CMOL circuits.

It is being proposed that relatively short and rigid molecules with about 100 atoms and two, or a few more, internal states would be the best choice for the development of molecular electronics [28]. If there were more internal states, the chance for a random change in the state might be too high.

One challenge here is the difficulty in writing proper software to reconfigure the chip. It is complex to design each configuration and also complex to design the software to do the reconfiguration.

Nanocells are an example of manufacturing with random or statistical features and then programming the end result as is needed. The nanocell is a molecular electronics design wherein a random self-assembled array of molecules and metallic nanoparticles are addressed by a relatively small number of input/output pins [20]. The cell is made by traditional lithographic means. In fabrication, the type and average density of molecules and metallic nanoparticles are controlled, but otherwise they land in random places on the cells. The result is a group of cells that one can program to follow Boolean logic. Circuits can be built by connecting the cells in a controlled way. Challenges with nanocells include the need for individual programming of each cell after fabrication, efficient construction of circuits from the nanocells, and the size of the nanocell being limited by the lithographic techniques.

Neural network architectures have received a lot of attention, due to their promise of low-power operation and intrinsic failure tolerance. Since the invention of formal neural networks, implementations of many neural machines have been proposed over the years, using all available technological resources, from analog and digital to optical. There have been some experiments in which artificial neurons have been used in CMOS, which have had some success for low-power, analog, low-bit sensory analysis [22]. Scaling to larger systems or general-purpose systems is, however, challenging. In CMOS, a limiting factor has been that physical implementation of artificial neurons has required a large area on the silicon surface, and the fact that compared with biological systems the connectivity of the neurons is low, which limits the performance. Also, there has been a lack of algorithms which are scalable to very large configurations; however, computational neuroscience is beginning to yield algorithms that can scale to large configurations and have the potential for solving large, very complex problems [18].

Now that nanotechnology has arrived on the scene, there may well be a revival of interest in these neural architectures. Indeed, many properties of this model seem particularly well suited to nanotechnology. Neural networks exhibit highly regular architectures. Like cellular automata, all the basic units are identical and carry out the same calculation. This is an eminently parallel architecture. The dynamics of neural networks is essentially asynchronous. Theoretically, therefore, there are no problems arising from distribution from a global clock. Neural networks are, in principle, very tolerant of errors. Harmful consequences of fabrication defects or operating errors are reduced, because the operating mode of these architectures results from the collective dynamics of distributed variables. As a consequence of the dynamics and the way they are programmed by learning, neural networks can be built from basic components with highly dispersed characteristics.

There are some difficulties due to which efficient nanoimplementations of neural networks architectures have not yet emerged. For certain network topologies, the interconnect circuit can become extremely complex, as can control of the associated weighting matrix. The computational power of neural networks lies in the large number of interconnects, without which it is difficult to compete with more general computation approaches. Also, the direct implantation of neural dynamics as proposed by the equations presented in Section 4.1 may make it difficult to implement the neuron and synaptic connections. A physical realization would require a summing device that can deal with a large amount of input, and a synapse function that can realize weighting functions during the relaxation phase and adaptation functions during the learning phase, i.e., multiplication operation or real numbers.

As a result, the neural network applications of cross-bar architectures discussed above are very limited due to the extremely simple types of molecular connections between the nanowires. The CMOL approach is an attempt to alleviate this by using CMOS cells to implement the more complex operations, but this leads to increased area and problems with interconnects between nano- and micro-scale devices.

Some ideas have been published concerning synapses made using molecular single-electron transistors [51] and adaptive synapses [52] capable of locally evaluating a learning rule like Hebb's rule. Although this work remains rather conceptual, it is expected that experimental demonstrations of neural nanodevices will follow soon, given the attractive aspect of the neural approach in the context of nanotechnology.

4.8   Memory technologies for the future

The dream of any system designer is to have unlimited amounts of fast memory. Already today mobile phones require a considerable amount of storage capacity to retain pictures, video, music, and a number of different applications. Taking into account different tools allowing a user to create his own content (multipixel still cameras, camcorders) and tools for fast loading of external content via fast data links (3G, WLAN, UWB, and USB) we can easily expect that mobile phones will require 50–100 GB internal mass memory.

However, fast memory is very expensive. The only economical solution for solving the dilemma between cost and performance is the use of different types of memory and organizing memory into several levels – each smaller, faster, and more expensive per byte than the next level lower, providing a memory system with cost as low as the cheapest level of memory and speed almost as fast as the fastest level. This is the reason why all modern mobile devices, like computers, use several types of memory.

The requirements for the memory elements of mobile devices are similar to those of the logic components:

- a low power consumption. Battery energy is limited and the maximum heat dissipation of products of the size of a mobile handheld device is about 2–4 W due to safety and reliability issues.
- a low voltage – related to power limitations, battery voltage development, and system design.
- a small form factor and low pin count: more functionality is needed in system ASICs but limited area and package/module height are requested.
- a low cost: the share of memories in total bill-of-material of the product is significant and constantly growing.

Mobile devices mainly use memories in which storage cells are organized in a matrix. Matrix-based memories can be grouped into read and write random access memory (RAM) and read only memory (ROM). In RAM, data can be written to or read from memory in any order, regardless of the memory location that was last accessed, i.e., at random. ROM can be divided into once-programmable ROM and reprogrammable ROM. Once-programmable ROM can be programmed during fabrication by design mask or by the customer, e.g., by fusing tiny metal bridges at the corresponding matrix nodes.

Reprogrammable ROM stores information in an array of memory cells that can be made from, e.g., floating-gate transistors. Each cell resembles a standard MOSFET, except that the transistor has two gates instead of one. On top is the control gate as in usual MOS transistors, but below this there is a floating gate insulated all around by an oxide layer. Because the floating gate is electrically isolated by its insulating layer, any charge placed on it is trapped there and, under normal conditions, will not discharge for many years. When the floating gate holds a charge, it screens the electric field from the control gate, which modifies the threshold voltage of the cell. The presence of charge within the gate shifts the transistor's threshold voltage, higher or lower, corresponding to a 1 to 0, for instance. Thus memory is nonvolatile because data are not lost even when the power is off.

Before new data are stored in a memory the old data must be erased by discharging the floating gates. For erasable programmable ROM (EPROM) this discharge is performed with ultraviolet light while for electrically erasable PROM (EEPROMs) it is done using enhanced programming voltage.

RAM devices are divided into volatile memories (VMs) and nonvolatile memories (NVMs). In VMs the information is lost after the power is switched off. NVMs retain the stored information even when not powered.

However, the border between nonvolatile RAM and reprogrammable ROM is very ill defined. Today, it is defined mainly by the write times, which are in the nanosecond–microsend range for nonvolatile RAM and higher for re-programmable ROM. Also, the term nonvolatile RAM is typically used for memory technologies based on novel materials which are not utilized in classical semiconductor technology.

The most important volatile memories are static RAM (SRAM) and dynamic RAM (DRAM).

SRAM consists of a series of transistors arranged in a flip-flop. SRAM exhibits data remanence. In this sense SRAM is static memory. But it is volatile because data are lost when power is off. Since the transistors have a very low power requirement, their switching time is very low. However, since an SRAM cell consists of several transistors, typically six or four, its density is quite low. SRAM is expensive, but very fast and not very power hungry. It is used only in small amounts when either speed or low power, or both, are important.

DRAM is a type of semiconductor memory that stores each bit of data in a separate capacitor. Since real capacitors leak charge, the information eventually fades unless the capacitor charge is refreshed periodically. Because of this refresh requirement, it is a dynamic memory unlike SRAM. The advantage of DRAM is its structural simplicity: only one transistor and a capacitor are required per bit (the so-called 1T-1C cell). This allows DRAM to reach very high density. DRAM is a volatile memory, since it loses its data when the power supply is removed. DRAM is power hungry. Several methods of lowering the power consumption of DRAM in mobile devices have been proposed, e.g., partial array self-refresh (only data in selected portion of memory are refreshed during self-refreshments), use of a temperature sensor to determine the refreshment rate, tighter leakage requirements, and so on [53].

4.8.1   Flash memory

The most important type of reprogrammable ROM is flash memory. Widely used industrial standards of flash architectures are NOR and NAND. Both NOR and NAND flash store information in an array of memory cells made from floating-gate transistors, but the cells are connected differently.

In NOR flash, cells are connected in parallel to the bit lines, allowing cells to be read and programmed individually. Reading from NOR flash is similar to reading from RAM. Because of this, most microprocessors can use NOR flash memory as execute in place

(XIP) memory, meaning that programs stored in NOR flash can be executed directly without the need to copy them into RAM. Although NOR can be read or programmed in a random access fashion, it must be erased a "block" at a time. In other words, NOR flash offers random-access read and programming operations, but cannot offer arbitrary random-access rewrite or erase operations. NOR-based flash has long erase and write times, but allows in random access to any memory location.

In NAND flash, cells are connected in series which prevents the cells being read and programmed individually. Because of the series connection, a large grid of NAND flash memory cells occupies only a small fraction of the area of equivalent NOR cells (assuming the same CMOS process resolution). This type of flash architecture combines higher storage capacity with faster erase, write, and read capabilities. However, the input/output interface of NAND flash does not provide a random-access external address bus. Rather, data must be read on a block-wise basis, with typical block sizes of hundreds to thousands of bits.

NAND flash is best suited to high-capacity data storage. NOR flash is targeted towards the program code or data storage applications. Because the share of data memory compared to code memory is increasing rapidly, the importance of NAND flash has been growing very quickly. Flash memory offers fast read access times and better kinetic shock resistance than hard disks. These characteristics explain the popularity of flash memory in portable devices.

In traditional single-level cell (SLC) flash, each cell stores only one bit of information. A newer flash memory, known as multilevel cell (MLC) devices, can store more than one bit per cell by choosing between multiple levels of electrical charge in the floating gates of its cells.

So far flash memory has been the most reliable mass storage technology for portable devices. However physical limitations for future flash cell scaling can already be seen: these include charge storage requirements of the dielectrics and reliability issues, capacitive coupling between adjustment cells, and high-voltage operation due to the energy barrier height [54].

Many innovations have been proposed to overcome the limitations in flash scaling using both the system management technique (error management techniques, multi-level memories) and improvements in fabrication technology (high-k dielectrics, nanocrystal or nitride traps for storage nodes, multi-layer barriers for tunnel dielectrics, Fin-FET, and three-dimensional architectures and so on) [54, 55].

4.8.2   Future options for memory technologies

While there is no doubt that flash will remain the dominant NVM technology at least down to the 32 nm node, the world is currently looking hard for an alternative. A range of new memory technologies have been explored: ferro-electric RAM (FeRAM), magnetic RAM (MRAM), polymer FeRAM, phase change memory (PCRAM), resistive RAM (RRAM), probe storage, CNT memory (NRAM), molecular and many others.

FeRAM is in some ways similar to DRAM, but uses a ferroelectric layer to achieve nonvolatility. In a FeRAM the cell capacitor is made of ferroelectric material,

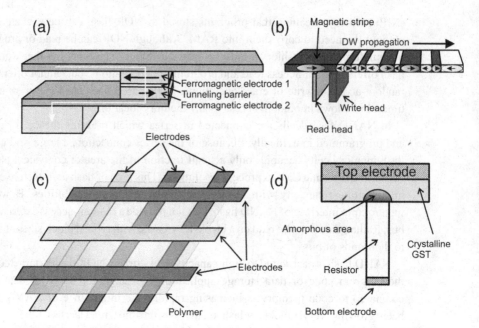

Figure 4.6 Several future options for memory technologies: (a) scheme of a MRAM cell; (b) RTM; (c) polymer memory; (d) phase change memory.

typically lead zirconate titanate (PZT). A ferroelectric material has a nonlinear relationship between the applied electric field and the polarization of the material. Characteristically with ferroelectric materials there is a hysteresis loop. When an external electric field is applied across a dielectric, the dipoles tend to align themselves with the field direction, produced by small shifts in the positions of atoms and shifts in the distributions of electronic charge in the crystal structure. After the charge is removed, the dipoles retain their polarization state. Binary "0"s and "1"s are stored as one of two possible electric polarizations in each data storage cell.

FeRAM is based on the physical movement of atoms in response to an external field, which happens to be extremely fast. In theory, this means that FeRAM could be much faster than DRAM. However, since power has to flow into the cell for reading and writing, the electrical and switching delays are likely to be similar to DRAM overall.

It is possible to make FeRAM cells using two additional masks during conventional CMOS semiconductor manufacturing. This makes FeRAM particularly attractive as an embedded NVM on microcontrollers, where the simpler process can reduce costs. However, the materials used to make FeRAMs are not commonly used elsewhere in integrated circuit manufacturing. FeRAM is competitive in specialized niche applications where its properties (lower power, fast write speed, and much greater write–erase endurance) give it a compelling advantage over flash memory.

In MRAM a memory cell is a magnetic storage element. The cell, as shown in Figure 4.6, is formed from two ferromagnetic films separated by a thin insulating layer. One of the two films has a high coercitivity thus fixing its magnetization in a particular direction. The second film has a lower coercitivity allowing its remagnetization by an external field. Due to the magnetic tunnel effect, the electrical resistance of the cell can be changed by altering the orientation of the magnetization in the films. Typically if the two films have the same polarity this is considered to mean "0", while if the two films are of opposite polarity the resistance will be higher and this means "1". A memory device is built from a grid of such "cells."

MRAM cells are placed at cross-points of an array of perpendicular conducting lines. In the conventional writing process current pulses are sent through two perpendicular lines and the resulting field is high enough to reverse the magnetization of the film with lower coercitivity. This approach requires a fairly substantial current to generate the field which makes it less interesting for low-power applications, which is one of MRAM's primary disadvantages. Additionally, as the device is scaled down in size, there comes a time when the induced field overlaps adjacent cells over a small area, leading to potential false writes. This problem appears to set a fairly large size for this type of cell.

MRAM has similar speeds to SRAM, similar density of DRAM but much lower power consumption than DRAM, and is much faster and suffers no degradation over time in comparison to flash memory. It is this combination of features that some suggest make it the "universal memory," able to replace SRAM, DRAM, and flash. This also explains the huge amount of research being carried out into developing it.

A new type of magnetic NVM so-called the *magnetic race-track memory* (RTM) was proposed by IBM [56]. The general principle of RTM is shown in Figure 4.6. The principle is reminiscent of the well-known magnetic tape storage. The information is stored in magnetization of domains in a magnetic stripe. However, instead of moving the stripe along read and write heads, the stripe is fixed above the heads. Magnetic domain walls are moved along the stripe (race track) by means of nanosecond long current pulses injected in the stripe. The direction of the pulse controls the direction of domain wall motion, and the quantity of current pulses is controlled to move the domain walls a specific number of bits. The domain walls are read and written with reading and writing heads. A three-dimensional version of RTM has been proposed in which there is a matrix of U-shaped nanowires standing perpendicularly on an Si chip above a two-dimensional plane of CMOS logic, reading, and writing elements. According to the IBM researchers a collection of these nanowires would provide a memory density higher than nearly all current solid-state offerings [57]. However, magnetic RTM is in the very early research phase and a lot of fundamental problems have to be solved before practical applications become possible.

The field of polymer electronics has emerged technically and commercially. Polymers have decisive advantages over silicon due to their minimal processing costs and the possibility of obtaining very different form factors (flexibility, moldability, transparency etc.). Several types of polymer memories have been proposed: polymer transistor RAM, polymer ferroelectric RAM (PFeRAM), polymer ionic RAM (PIRAM) [58, 59, 60].

In PFeRAM a substrate is coated with thin layers of ferroelectric polymer. The polymer layer is placed between two sets of crossed electrodes. Data are stored by changing the polarization of the polymer between the electrodes. Each point of intersection represents a memory cell containing one bit of information. Ultrathin layers of polymers can be stacked, so expanding the memory capacity is simply a matter of coating a new layer of polymer on top of an existing one. Thus these types of approaches can provide truly three-dimensional memory architecture. The structure of a PIRAM cell is similar to that of a PFeRHM cell. Application of an electric field to a cell lowers the polymer's resistance; the polymer maintains its state until a field of opposite polarity is applied to raise its resistance back to its original level [60]. The different conductivity states represent bits of information.

The combination of a low-cost fabrication process and the possibility of three-dimensional architecture in principle offer the prospect of cheap memory with a reasonable capacity. However, the main challenge lies in creating suitable polymer materials. It seems this task is much more difficult than originally thought.

Phase change memory, PCRAM, uses the unique behavior of chalcogenide glass, which can be "switched" between two states, crystalline and amorphous, with the application of heat. The crystalline and amorphous states of chalcogenide glass have dramatically different electrical resistivities, and this forms the basis by which data are stored. The majority of PCRAM devices use a chalcogenide alloy of germanium, antimony, and tellurium (GeSbTe) called GST. It is heated to a high temperature (over 600 °C), at which point the chalcogenide becomes a liquid. Once cooled, it is frozen into an amorphous glass-like state and its electrical resistance is high. Heating the chalcogenide to a temperature above its crystallization point, but below the melting point, transforms it into a crystalline state with a much lower resistance. This phase transition process can be completed in just few nanoseconds.

A more recent advance allows the material state to be more carefully controlled, so that it can be transformed into one of four distinct states, namely the previous amorphous and crystalline states, and two new partially crystalline ones [61]. Each of these states has different electrical properties that can be measured during reads, allowing a single cell to represent two bits, thus doubling the memory density.

PCRAM can offer much higher performance in applications where writing quickly is important, both because the memory element can be switched more quickly, and also because single bits may be changed to either 1 or 0 without needing to first erase an entire block of cells.

The greatest challenge for PCM has been the requirement of high programming current density in the active volume. This has led to active areas which are much smaller than the driving transistor area. The discrepancy has forced phase-change memory manufacturers to package the heater and sometimes the phase-change material itself in sublithographic dimensions. The contact between the hot phase-change region and the adjacent dielectric is another fundamental concern. The dielectric may begin to leak current at higher temperature, or may lose adhesion when expanding at a different rate from the phase-change material. However, there has been a lot of progress in the development of PCM. PCM shows excellent reliability (10 years), very high endurance (up to 10^{12} cycles), and a very great potential for scaling.

In RRAM, the so-called colossal electroresistive (CER) resistor is used as a memory element [62]. The resistance of the resistor can be changed to high resistance by using a narrow electric pulse. Changed resistance is stable, which means it is nonvolatile.

Cantilever array on a CMOS chip

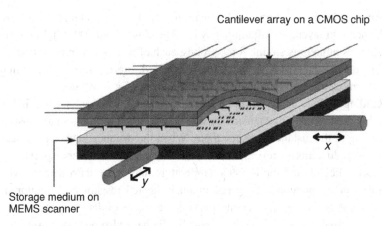

Storage medium on
MEMS scanner

Figure 4.7 Schematic of a probe storage memory.

The memory can be reset to a low resistance state by applying opposite polarity pulses. The RRAM memory cell can be also programmed to either a high resistance state or a low resistance state by using multiple pulses with relatively low amplitude. Therefore, multilevel cells are possible.

The basic principal of RRAM is that a dielectric, which is normally insulating, can be made to conduct through a filament or conduction path formed after application of a sufficiently high voltage. The conduction path formation can arise from different mechanisms, including defects, metal migration, charge trapping near the electrode interface etc. Once the filament is formed, it may be reset (broken, resulting in high resistance) or set (reformed, resulting in lower resistance) by an appropriately applied voltage. Different forms of RRAM have been proposed, based on different dielectric materials, ranging from perovskites to transition metal oxides to chalcogenides [63].

RRAM exhibits low programming power without sacrificing programming speed, retention, or endurance.

The *probe storage* concept shown in Figure 4.7 is based on a mechanical parallel *x/y* scanning of either the array of probes or the storage medium. Each probe is responsible for reading and writing data in its own small storage field. The medium is moved to position the selected bits under the probe using electromechanical actuators. In addition, a feedback-controlled *z*-approaching scheme brings the probe array into contact with the storage medium. This contact is maintained and controlled while *x/y* scanning is performed for write/read. Different read/write mechanisms (magnetic, resistive, piezoelectric, atomic force probes) and different storage media (magnetic, ferroelectric, phase-change, thermomechanical) have been proposed and demonstrated in probe storage devices [64].

So far IBM has made the greatest progress in the development of probe storage technology. It was not only the first to propose the concept of probe memory itself but it has been able to demonstrate key subsystems and finally complete a probe storage prototype, the so-called Millipede [65]. The Millipede storage approach is based on a thermomechanical write and read processes in nanometer-thick polymer films.

Thermomechanical writing is a combination of applying a local force by the cantilever/tip to the polymer layer and softening it by local heating (about 400 °C). Mechanically, Millipede uses numerous atomic force probes, each of which is responsible for reading and writing data in the area with which it is associated. Bits are stored as a pit in the surface of a thermoactive polymer deposited as a thin film on a substrate.

Reading is done using thermomechanical sensing. Thermal sensing is based on the fact that the thermal conductance between the heater and the storage substrate changes according to the distance between them. When the distance between the heater and the sample is reduced as the tip moves into a bit indentation, the heat transport through air will be more efficient, and the heater's temperature and hence its resistance will decrease. Changes in temperature of the continuously heated resistor are monitored while the cantilever is scanned over data bits, providing a means of detecting the bits. Erasing and rewriting capabilities of polymer storage media have also been demonstrated. Thermal reflow of storage fields is achieved by heating the medium to about 150 °C for a few seconds. The smoothness of the reflowed medium allowed multiple rewriting of the same storage field.

Heating the probes requires quite a lot of power for general operation. However, the exact power consumption depends on the speed at which data are being accessed. A higher data rate requires higher power consumption.

The probe storage technique is capable of achieving data densities exceeding 1 Tb/in^2, well beyond the expected limits of magnetic recording. However the development of probe storage into a commercially useful product has been slower than expected because the huge advances in flash and hard drive technologies.

The nanotube RAM (NRAM) uses nanotubes as electromechanical switches to represent data [66]. Nanotubes are suspended above a metal electrode. This represents a "0" state. If the nanotubes are charged with positive charge and the electrode with negative charge, the nanotubes bend down to touch the metal electrode. This represents a "1" state. The nanotubes then bind to the electrode by van der Waals forces and stay there permanently. It is therefore nonvolatile. The state of the electromechanical switch can be changed by reversing the polarity of electrical field. The position of the nanotubes can be read out as a junction resistance. Nanotube motion happens extremely fast because of the nano-size of the switch and also uses very little power. In theory NRAM can provide a good combination of properties: high speed (comparable to SRAM), small cell size (similar to DRAM), unlimited lifetime (>10^{15} cycles), resistance to heat, cold, magnetism, and radiation. However, so far the demonstrated parameters of NRAM are still very far from theoretical predictions.

Molecular memory utilizes molecules as the data storage element. In an ideal molecular memory device, each individual molecule contains a bit of data, leading to a very high storage capacity. However, so far in practice devices are more likely to utilize large numbers of molecules for each bit. The molecular component can be described as a molecular switch, and may perform this function by different mechanisms, including charge storage, photochromism, or changes in capacitance.

ZettaCore, Inc. have proposed molecular memory based on the properties of specially designed molecules [67]. These molecules are used to store information by adding or removing electrons and then detecting the charge state of the molecule. The molecules, called multiporphyrin nanostructures, can be oxidized and reduced (electrons removed or added) in a way that is stable, reproducible, and reversible. In many ways, each molecule acts like an individual capacitor device, similar to a conventional capacitor, but storing only a few electrons of charge that is accessible only at specific, quantized voltage levels. Using multistate molecular memory with quantized energy states, more than one bit of information can be stored in a single memory location [68].

Hewlett-Packard have developed a molecular switch in which the active element is made of a single layer of Rotaxane molecules containing several million molecules sandwiched between bottom and top electrodes [69]. It has also announced that it has fabricated a 64-bit array that switches molecules on and off inside a grid of nanowires only a square micrometer in size [70]. However, molecules in the prototype could only be turned on and off a few hundred times before degrading.

The potential benefits of molecular memory are high density and fast access. However, it is still in the very early stage of research and further fundamental work is needed to create a knowledge base for molecular electronics.

4.8.3   Comparison of the memory technologies

All of the mentioned memory technologies have different levels of maturity from well-known technologies that are manufactured commercially to those that are under development and even those in the first stages of research. In Table 4.1 we briefly summarize the current status of some of the possible future memory technologies and estimated parameters, based on [24].

Taking into consideration requirements for mass storage used in portable devices, physical limitations, and the current status of technologies PCRAM looks the most promising. However, a lot of work needs to be done before it will be able to compete with flash.

4.9   Comparisons of nanocomputing solutions for mobile devices

In order to select the right kind of implementation for computing for a future mobile device, we need a quantitative evaluation of the performance and cost of the different options. There are many measures of performance, and the targeted end application determines the relevance of each performance metric.

Full quantitative evaluation of the merits of various nanocomputing possibilities is at the moment impossible due to missing data. But there are nevertheless some conclusions and comparisons that can be made even today. The main driver behind nanocomputing is to find new domains for scaling information processing functional density and throughput per joule to substantially beyond the values attainable with developing silicon CMOS. But there are also other requirements which can be more relevant, keeping in mind the particular applications and uses. The most important comparison criteria for mobile devices are scalability, support for integration, overall suitability to the desired

Table 4.1. Memory technologies: demonstrated and projected parameters; F is technology mode; minimum geometries producible in IC manufacturing

Storage mechanism				Memory technologies				
		Charge	Charge	Electrical polarization	Magnetization	Phase changing	Nanomechanical switch	Multiple mechanisms
Parameters	Year	NOR Flash	NAND Flash	FeRAM	MRAM	PCRAM	NRAM	RRAM
Feature size F, nm	2007	90	90	180	90	65	180	90
	2022	18	18	65	22	18	5–10	5–10
Cell area	2007	$10F^2$	$5F^2$	$22F^2$	$20F^2$	$4.8F^2$	No data	$8F^2$
	2022	$10F^2$	$5F^2$	$12F^2$	$16F^2$	$4.7F^2$	$5F^2$	$8/5F^2$
Read time	2007	10 ns	50 ns	45 ns	20 ns	60 ns	3 ns	No data
	2022	2 ns	10 ns	<20 ns	<0.5 ns	<60 ns	3 ns	10 ns
Write/erase time	2007	1 µs/10 ms	1 ms/0.1 ms	10 ns	20 ns	50/120 ns	3 ns	100 ns
	2022	1 µs/10 ms	1 ms/0.1 ms	1 ns	<0.5 ns	<50 ns	3 ns	<20 ns
Retention time, year	2007	>10	>10	>10	>10	>10	Days	>10
	2022	>10	>10	>10	>10	>10	>10	>10
Write cycles	2007	$>10^5$	$>10^5$	10^{14}	$>10^{16}$	10^8	$>10^9$	$>10^5$
	2022	$>10^5$	$>10^5$	10^{16}	$>3 \times 10^{16}$	10^{15}	$>3 \times 10^{16}$	$>10^{16}$
MLC operation		Yes	Yes	No	No	Yes	No	Yes

task, manufacturability and competitive price, low power consumption per unit operation, high enough speed, high memory density, flexibility in physical form, reliability, and environmental aspects.

For quite some time to come, manufacturability will set the most severe constraints on the competitiveness of the different approaches. Nanotechnology has been enabled by new techniques for building and measuring structures at the nanoscale, but many of these techniques are still only suitable for laboratory work. In order to be relevant for future mobile devices, the new inventions have to be suitable for mass manufacturing. In addition, the manufacturing methods will have to be as simple as possible in order to ensure cost effectiveness. In many cases compatibility with CMOS technologies will be required for integration with current technologies and also to lower the costs by using well-known methods and tools.

At the nanoscale the operation of the devices is more stochastic in nature and mesoscopical phenomena like quantum effects become the rule rather than the exception. Even though this is a great opportunity, reduced dimensions will result in more failing or undeterministic components than in traditional electronics. Therefore, strategies are needed for designs that are tolerant to failing components, and capable of taking into account the quantum-scale effects inherent in nanosystems. As discussed, there are two ways to achieve this: either by using redundant hardware, or by using software. But in

Table 4.2. Comparisons between different nanoscale architectures and/or computing logic principles

	Von Neumann and digital logic	Reconfigurable logic	Neuromimetic or cellular non-linear networks	Cellular automata
Application	Generic computation	Generic computation	Sensory data, classification	Sensory data, classification
Failure tolerance	Software implementation	Specifically design circuits	Intrinsic if full parallel circuits	Good/depends on the rule
Dynamic change	Upload new software	Dynamical partial reconfiguration triggered by software	On-line learning algorithm	Upload new CA rule
Volatility of resources	Software implemented thru publish/discover protocol	Hardware/Software implementation of publish/discover protocol	Connect/Disconnect of sub-networks plus on-line learning	?
Clocking	Needed	Needed	Not needed	Not needed; depends on implementation
Merits	Generality, established design tools exits	Tolerance to failures	Robustness, speed	Robustness, speed
Dismerits	Failure tolerance, manufacturing in nanoscale difficult, memory access	Inefficient use of area, manufacturing	Application specific, implementation of learning algorithms or teaching the system	Application specific

order to design a practical realizable nanocomputing device it is necessary to balance the ease of manufacturing with the ease of programming, and the right balance will depend on the chosen application.

For most of the other attributes, comparisons are quite difficult to make – due not only to missing data but also to the fact that not all criteria are relevant for all computing principles or architectures. For example, clock speed and switching energy are irrelevant when the target application is based on analog processing, but are crucial for comparing different digital devices. Noise tolerance depends on the physical realization and the computing logic used, etc. Tables 4.2 and 4.3 summarize the various computer architecture alternatives proposed for nanocomputing systems and their possible physical implementation technologies in regard to the basic requirements for mobile computing. In Table 4.2 the architecture also includes elements from the computational principle, and implementation technology combines the physical basis and the components used. Table 4.3 lists some of the possible implementation alternatives for the architectures. The numbers in the table are projected estimates from [23, 24]. All the technologies in the table are based on electromagnetism, because they are more likely to be relevant

Table 4.3. Comparisons of the physical implementation alternatives

	CMOS FET	Carbon-based FETs	Spintronic FET	Molecular FETs	Magnetic logic
Manufacturability	Lithography works well up to certain densities, difficult to scale down from that	Yield, process variability is large	Process compatibility and variability	Yield, process variability, chemical assembly good for mass manufacturing	Regular structures, easy to manufacture
Potential competitiveness	Benchmark	Material cost	Not clear yet	High density	Low power operation, keep the state when power is off – no need for memory
CMOS compatibility	Automatic	Can be good, depends on implementation	Can be good, depends on implementation	Can be good, depends on implementation	Can be good, depends on implementation
Input and output	Easy	Challenging	Good for CMOS compatible implementation	Challenging	Challenging
Interconnects	Easy	Challenging	Good for CMOS compatible implementation	Challenging	Challenging
Switch energy J	3E-18	3E-18	3E-18	5E-17	Lower than in the others
Switching time, speed	12 THz	6.3 THz	40 GHz	1 THz	1 GHz
Density (units per cm^2)	1E10	5E9	5E9	1E12	5E9
Error tolerance	Bad with current method of use	To be studied, depends on architecture	To be studied, depends on architecture	To be studied, depends on architecture	Good?
Noise tolerance	Good up to a limit	To be studied, depends on architecture	To be studied, depends on architecture	To be studied, depends on architecture	Good?
Gain[a]	Good	Challenging for other solutions than a transistor	Challenging for other solutions than a transistor	Challenging for other solutions than a transistor	Not needed
Material used	Silicon	Carbon	Si, III–V, metaloxides	Organic molecules	Ferromagnetic materials
Aging and reliability	8% more vulnerable to radiation per each generation (Intel presentation in IWFIPT07)	Not known	Not known	Not known	Tolerant to aging? Good radiation tolerance

[a] According to [23], gain of nanodevices is an important limitation for current combinatorial logic where gate fan-outs require significant drive current and low voltages make gates more noise sensitive.

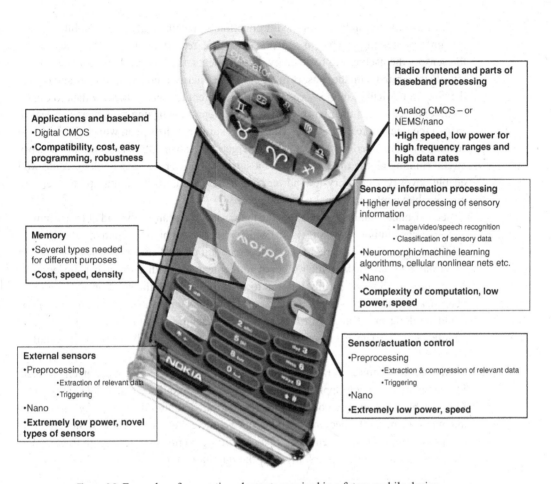

Radio frontend and parts of baseband processing

•Analog CMOS – or NEMS/nano
•High speed, low power for high frequency ranges and high data rates

Applications and baseband
•Digital CMOS
•Compatibility, cost, easy programming, robustness

Sensory information processing
•Higher level processing of sensory information
    • Image/video/speech recognition
    • Classification of sensory data
•Neuromorphic/machine learning algorithms, cellular nonlinear nets etc.
•Nano
•Complexity of computation, low power, speed

Memory
•Several types needed for different purposes
•Cost, speed, density

External sensors
•Preprocessing
    •Extraction of relevant data
    •Triggering
•Nano
•Extremely low power, novel types of sensors

Sensor/actuation control
•Preprocessing
    •Extraction & compression of relevant data
    •Triggering
•Nano
•Extremely low power, speed

Figure 4.8 Examples of computing elements required in a future mobile device.

for future mobile devices, due to easier interoperability with current technologies. Also, only implementation alternatives which can operate at room temperature were included.

Looking at the tables it is clear that there is no easy winner over CMOS when it comes to general computing. Spintronics is very promising in this respect, but there are still many problems to be solved. Logic with magnetic nanoparticles and cellular automata could become a competitive option in some application areas needing low-power computing. Of course, there are many problems to be overcome here as well, but low power is a good motivation from the point of view of mobile devices and intelligence embedded in the environment.

However, as discussed in the introduction, it is likely that future mobile devices will host different types of computational and signal processing units targeted at different purposes. A possible scenario for a future mobile device is presented in Figure 4.8. Digital CMOS processors will dominate the computing needed for applications and parts of the radio baseband processing for some time to come, due to compatibility with existing software, easy programmability, cost, and robustness.

The most likely candidate for the introduction of nanotechnologies in mobile devices is signal processing related to sensory data. The reason for this is twofold: first if the sensors themselves are realized by nanoelements, the low-level control needs to be closely integrated with the sensors in order to save energy, increase the speed, and reduce the size. In particular, in many cases it is not efficient to convert the raw data to digital form for processing, rather it is better to build a preprocessing unit from nanoelements to extract the relevant features of the data. This is similar to the way in which all biological sensory systems operate. Also, it is not energetically favorable to keep the full sensory circuit running all the time, waiting for a signal to register. Nanotechnology could well enable novel types of very-low-power triggering circuits for "waking up" the sensory circuit when a relevant signal will be detected.

Second, understanding and classifying the sensory data can easily lead to very complex and time-consuming processing tasks, which require very high speed in order to be truly useful. Applications like real-time video recognition and understanding the context of visual data are tasks which are not implementable with current mobile devices. Here, nanocircuits specifically tailored to the task could be the solution. Also simpler classifications related to smell, touch, voice, or radio signal sensing for cognitive radios could be done faster and with less energy by machine learning circuits made using nanotechnology. The many possibilities of nanosensing will be covered in detail in Chapter 5.

Also certain features of radio signal processing could benefit from nanotechnological implementation. When approaching high operating frequencies and data rates, tasks like synchronization, (de)coding, and modulation for multiple frequency and multi-antenna transmissions become challenging. Nanoelectromechanical systems could be one possibility for fast and energy efficient radio signal processing for high frequencies, but there are still severe problems with the quality of the components. Radio aspects will be discussed in detail in Chapter 6.

Unfortunately, there are still no clear winners in the nanocomputing race, even for application-specific computing tasks. There are severe difficulties with architecture and manufacturing: so far none of the proposals has been able to combine successfully the requirements for easy manufacturing, efficient computing, good compatibility with current mainstream technologies, and sufficiently high reliability. Even with the memory devices, which are simpler than any information processing systems, no successful solutions capable of competing with current technologies have yet emerged. But, hopefully, in the near future we will see how new nanocomputing systems evolve from thought experiments to successful laboratory demonstrations, and finally to mass manufacturing. The last stage is where the battles for winning solutions will be fought, and the one that will prove to be the toughest challenge.

Acknowledgments

We thank Asta Kärkkäinen and Jaakko Peltonen for background material and the Finnish Funding Agency for Technology and Innovation TEKES for financial support.

References

[1] A. Abraham, C. Grosan, and V. Ramos, eds., *Swarm Intelligence in Data Mining*, p. 270, Springer Verlag, 2006.

[2] L. Adleman, Molecular computation of solutions to combinatorial problems. *Science*, **266**, 1021–1024, 1994.

[3] P. Beckett and A. Jennings, Towards nanocomputer architecture, in *Proceedings of the Seventh Asia-Pacific Computer Systems Architecture Conference (ACSAC 2002)*, F. Lai and J. Morris, eds., ACS, 2002.

[4] G. Bourianoff, The future of nanocomputing, *Computer*, **36**, no. 8, 44–53, 2003.

[5] M. Branicky, *Analog Computation with Continuous ODEs*, IEEE, 1994.

[6] R. W. Brockett, Input–output properties of the Toda-lattice, in *Proceedings of IEEE 43rd Conference on Decision and Control (2004)*, IEEE, 2004.

[7] R. W. Brockett, Dynamical system that sort lists, diagonalize matrices and solve linear programming problems, in *Proceedings of the IEEE Conference on Decision and Control (1988)*, IEEE, 1988.

[8] R. W. Brockett, Dynamical systems and their associated automata, in *Systems and Networks: Mathematical Theory and Applications*, U. Helmke *et al.*, eds. Akademie Verlag, 1994, and R. W. Brockett, Smooth dynamical systems which realize arithmetical and logical operations, in *Three Decades of Mathematical Systems Theory*, H. Nijmeijer and J. M. Schumacher, eds., pp. 19–30, Lecture Notes in Control and Information Sciences, Springer-Verlag, 1984.

[9] R. P. Cowburn and M. E. Welland, Room temperature magnetic quantum cellular automata, Science, **287**, 1466, 2000.

[10] M. I. Dyakonov and M. Shur, Plasma wave electronics: novel terahertz devices using two dimensional electron fluid, *IEEE Trans. Electron. Devices*, **43**, no. 10, 1996.

[11] A. Engelbrecht, *Fundamentals of Computational Swarm Intelligence*. Wiley & Sons, 2005.

[12] M. P. Frank, Course on Physical Limits of Computing, http://www.cise.ufl.edu/~mpf/physlim/.

[13] M. P. Frank, Nanocomputer systems engineering, in *Technical Proceedings of the 2003 Nanotechnology Conference and Trade Show*, vol. 2, pp. 182–185, NSTI, 2003.

[14] M. P. Frank and T. Knight, The ultimate theoretical models of nanocomputers, *Nanotech.*, **9**, 162–176, 1998.

[15] S. Fujita, K. Nomura, K. Abe, and T. H. Lee, Novel architecture based on floating gate CNT-NEMS switches and its application to 3D on-chip bus beyond CMOS architecture, in *Proceedings of the Sixth IEEE Conference on Nanotechnology (IEEE-NANO 2006)*, vol. 1, pp. 17–20, 314, IEEE, 2006.

[16] S. Fujita, K. Nomura, K. Abe, and T. H. Lee, 3-D nanoarchitectures with carbon nanotube mechanical switches for future on-chip network beyond CMOS architecture, *IEEE Trans. Circuits and Systems I: Regular Papers*, **54**, no. 11, 2472, 2007.

[17] M. Greenstreet and X. Huang, A smooth dynamical system that counts in binary, *Proceedings of the IEEE International Symposium on Circuits and Systems (1997)*, IEEE, 1997.

[18] D. Hammerstrom, A survey of bio-inspired and other alternative architectures, in *Nanotechnology*, vol. 4: *Information Technology*, R. Waser, ed., pp. 251–285, Wiley-VCH, 2008.

[19] S. Haque, M. Yamamoto, R. Nakatani, and Y. Endo, Magnetic logic gate for binary computing, *Sci. Tech. Adv. Mat.*, **5**, 79, 2004.

[20] P. C. Husband, S. M. Husband, S. Daniels, and J. M. Tour, Logic and memory with nanocell circuits. *IEEE Trans. Electron. Devices*, **50**, no. 9, 1865–1875, 2003.

[21] A. Imre, G. Csaba, L. Ji, *et al.*, Majority logic gate for magnetic quantum-dot cellular automata, *Science*, **311**, no. 5758, 205–208, 2006.

[22] G. Indiveri and R. Douglas, Neuromorphic networks of spiking neurons. In *Nano and Molecular Electronics Handbook*, S. E. Lyshevski, ed., CRC Press, 2007.

[23] ERD Working Group, *International Technology Roadmap for Semiconductors*, ITRS Roadmap Emerging Research Devices, ITRS, 2005.

[24] ERD Working Group, *International Technology Roadmap for Semiconductors*, ITRS Roadmap Emerging Research Devices, ITRS, 2007; http://www.itrs.net/Links/2007ITRS/Home2007.htm.

[25] A. Khitun and K. L. Wang, Nano logic circuits with spin wave bus, in *Proceedings of the Third International Conference on Information Technology: New Generations* (ITNG'06), IEEE, 2006.

[26] Y. Kuramoto, in *International Symposium on Mathematical Problems in Theoretical Physics*, H. Araki, ed., Lecture Notes in Physics, No 30, Springer, 1975.

[27] C. S. Lent and B. Isaksen, Clocked molecular quantum-dot cellular automata. *IEEE Trans. Electron. Devices*, **50**, no. 9, 1890–1896, 2003.

[28] K. K. Likharev and D. B. Strukov, CMOL: devices, circuits, and architectures, in *Introducing Molecular Electronics*, G. Cuniberti, G. Fagas, and K. Richter, eds., Lecture Notes in Physics, Springer, 2005.

[29] K. K. Likharev, Hybrid semiconductor/Nan electronic circuits: freeing advanced lithography from the alignment accuracy burden, *J. Vac. Sci. Technol.* B. **25**, no. 6, 2531–2536, 2007.

[30] R. Blick, H. Qin, H.-S. Kim and R. Marsland, A nanomechanical computer – exploring new avenues of computing, *New J. Phys.*, **9**, 241, 2007.

[31] D. J. Monsma, J. C. Lodder, T. J. A. Popma, and B. Dieny, Perpendicular hot electron spin-valve effect in a new magnetic field sensor: the spin-valve transistor, *Phys. Rev. Lett.*, **74**, 5260, 1995.

[32] K. Nikolic, M. Forshaw, and R. Compano, The current status of nanoelectronic devices. *Int. J. Nanoscience*, **2**, 7–29, 2003.

[33] E. Ozbay Plasmonics: merging photonics and electronics at nanoscale dimensions. *Science* **311**, 5758, 189–193, 2006.

[34] S. Reitzenstein, L. Worschech, P. Hartmann, and A. Forchel, Logic AND/NAND gates based on three-terminal ballistic junctions, *Electron. Lett.*, **38**, no. 17, 951, 2002.

[35] R. Sarpeshkar, Analog versus digital: extrapolating from electronics to neurobiology, *Neural Computation*, **10**, 1601–1638, 1998.

[36] G. Seelig, D. Soloveichik, D. Y. Zhang, and E. Winfree, Enzyme-free nucleic acid logic circuits. *Science*, **314**, 1585–1588, 2006.

[37] M. R. Stan, P. D. Franzon, S. C. Goldstein, J. C. Lach, and M. M. Ziegler, Molecular electronics: from devices and interconnects to circuits and architecture. *Proc. IEEE*, **91**, no. 11, 1940–1957, 2003.

[38] K. Terabe, T. Hasegawa, T. Nakayama, and M. Aono, Quantized conductance atomic switch, *Nature*, **433**, 47, 2005.

[39] M. A. Uusitalo, and J. Peltonen, Nanocomputing with machine learning, Nanotech Northern Europe NTNE, Copenhagen, 2008.

[40] B. Vigoda, Analog logic: continuous-time analog circuits for statistical signal processing, PhD Thesis, MIT, 2003.

[41] D. Vion, A. Aassime, A. Cottet, *et al.*, Manipulating the quantum state of an electrical circuit, *Science*, **296**, no. 5569, 886–889, 2002.

[42] S. Wolf, D. D. Awschalom, R. A. Buhrman, *et al.*, Spintronics: a spin-based electronics vision for the future, *Science*, **294**, 1488, 2001.

[43] Q. Xu and M. Lipson, All-optical logic based on silicon micro-ring resonators, *Optics Express*, **15**, no. 3, 924, 2007.

[44] V. V. Zhirnov, R. K. Cavin, J. A. Hutchby, and G. I. Bourianoff, Limits to binary logic switch scaling – a gedanken model, *Proc. IEEE*, **91**, no. 11, 1934–1939, 2003.

[45] T. Toffoli and N. Margolus, *Cellular Automata Machines: A New Environment for Modeling*, MIT Press, 1987.

[46] M. Davis, *The Universal Computer: The Road from Leibniz to Turing*, W. W. Norton & Company, 2000.

[47] W. Aspray, *John Von Neumann and the Origins of Modern Computing*, MIT Press, 1991.

[48] J. R. Heath, P. J. Kuekes, G. S. Snider, and R. S. Williams, A defect-tolerant computer architecture: opportunities for nanotechnology, *Science*, **280**, 1716–1721, 1998.

[49] J. J. Hopfield, Neural networks and physical systems with emergent collective computational abilities, *PNAS*, **79**, 2554–2558, 1982.

[50] F. Rosenblatt, The perceptron: a probabilistic model for information storage and organization in the brain, *Psychological Rev.*, **65**, 386–408, 1958.

[51] S. Fölling, Ö. Türel, and K. K. Likharev, Single-electron latching switches as nanoscale synapses, in *IJCNN'01 Neural Networks (2001)*, pp. 216–221, IEEE, 2001.

[52] Q. Lai, Z. Li, L. Zhang, *et al.*, An organic/Si nanowire hybrid field configurable transistor, *Nano Lett.*, **8**, 876–880, 2008.

[53] F. Morishita, I. Hayashi, H. Matsuoka, *et al.*, A 312-MHz 16-Mb random-cycle embedded DRAM macro with a power-down data retention mode for mobile applications, *IEEE J. Solid-State Circuits*, **40**, no. 1, 204–212, 2005.

[54] S. K. Lai, Flash memories: successes and challenges, *IBM J. Res. Dev.*, **52**, N4/5, 529–553, 2008.

[55] G. W. Burr, B. N. Kurdi, J. C. Scott, C. H. Lam, K. Gopalakrishnan, and R. S. Shenoy, Overview of candidate device technologies for storage-class memory, *IBM J. Res. Dev.*, **52**, N4/5, 449–463, 2008.

[56] S. S. P. Parkin, Spintronic materials and devices: past, present and future!, in *Proceedings of the IEEE International, Electron Devices Meeting (2004)*, pp. 903–906, IEEE, 2004.

[57] S. S. P. Parkin, M. Hayashi, and L. Thomas, Magnetic domain-wall racetrack memory, *Science*, **320**, no. 5873, 190–194, 2008.

[58] C. J. Drury, C. M. J. Mutsaers, C. M. Hart, M. Matters, and D. M. de Leeuw, Low-cost all-polymer integrated circuits, *Appl. Phys. Lett.*, **73**, 108, 1998.

[59] www.thinfilm.se

[60] J. C. Scott and L. D. Bozano, Nonvolatile memory elements based on organic materials, *Adv. Mater.*, **19**, 1452–1463, 2007.

[61] F. Bedeschi, R. Fackenthal, C. Resta, *et al.*, A multi-level-cell bipolar-selected phase-change memory, in *Proceedings of the Solid-State Circiuts Conference (2008)*, pp. 428–429, IEEE, 2008.

[62] W. W. Zhuang, W. Pan, B. D. Ulrich, *et al.*, Novel colossal magnetoresistive thin film nonvolatile resistance random access memory (RRAM), in *Electron Devices Meeting (IEDM '02) Digest*, IEEE, 2002.

[63] G. I. Meijer, Who wins the nonvolatile memory race, *Science*, **319**, 1625–1626, 2008.

[64] S. Hong and N. Park, Resistive probe storage: read/write mechanism, in *Scanning Probe Microscopy*, Springer, 2007.

[65] P. Vettiger, M. Despont, U. Drechsler, *et al.*, The "Millipede"-More than one thousand tips for future AFM data storage, *IBM J. Res. Dev.*, 2000.

[66] T. Rueckes, K. Kim, E. Joselevich, *et al.*, Carbon nanotube-based nonvolatile random access memory for molecular computing, *Science*, **289**, 94–98, 2000.

[67] Z. Liu, A. A. Yasseri, J. S. Lindsey, and D. F. Bocian, Molecular memories that survive silicon device processing and real-world operation, *Science*, **302**, no. 5650, 1543–1545, 2003.

[68] L. Wei, K. Padmaja, W. J. Youngblood, *et al.*, Diverse redox-active molecules bearing identical thiol-terminated tripodal tethers for studies of molecular information storage, *Org. Chem.*, **69**, no. 5, 1461–1469, 2004.

[69] C. P. Collier, E. W. Wong, M. Belohradsky, *et al.*, Electronically configurable molecular-based logic gates, *Science*, **285**, 331, 1999.

[70] E. Smalley, HP maps molecular memory, *TRN News*, July 18, 2001.

[71] N. Nefedov, Spectral sensing with coupled nanoscale oscillators, in *Proceedings of the Second International Conference on Nano-Networks (2007)*, ICST, 2007.

[72] W. Wang and C. Moraga, Design of multivalued circuits using genetic algorithms, in *Proceedings of the 26th International Symposium on Multiple-Valued Logic*, pp. 216–221, IEEE, 1996.

[73] See, e.g., C. M. Bishop, *Pattern Recognition and Machine Learning*, Springer, 2006, or E. Alpaydin, *Introduction to Machine Learning*, MIT Press, 2004.

[74] See, e.g., http://www.notaworld.org.

[75] J. Peltonen, M. A. Uusitalo, and J. Pajarinen, Nano-scale fault tolerant machine learning for cognitive radio, in *IEEE Workshop on Machine Learning for Signal Processing Cancun (2008)*, pp. 163–168, IEEE, 2008.

5 Sensing, actuation, and interaction

P. Andrew, M. J. A. Bailey, T. Ryhänen, and D. Wei

5.1 Introduction

5.1.1 Ubiquitous sensing, actuation, and interaction

The London of 2020, as described in Chapter 1, will have conserved most of its old character but it will also have become a mixed reality built upon the connections between the ubiquitous Internet and the physical world. These connections will be made by a variety of different intelligent embedded devices. Networks of distributed sensors and actuators together with their computing and communication capabilities will have spread throughout the infrastructures of cities and to various smaller objects in the everyday environment. Mobile devices will connect their users to this local sensory information and these smart environments. In this context, the mobile device will be a gateway connecting the local physical environment of its user to the specific digital services of interest, creating an experience of mixed virtual and physical realities. (See also Figure 1.1.)

Human interaction with this mixed reality will be based on various devices that make the immediate environment sensitive and responsive to the person in contact with it. Intelligence will become distributed across this heterogeneous network of devices that vary from passive radio frequency identification (RFID) tags to powerful computers and mobile devices. In addition, this device network will be capable of sharing information that is both measured by and stored in it, and of processing and evaluating the information on various levels. The networks of the future will be cognitive systems that consist of processes capable of: (1) perception of the physical world, through various sensory processes, (2) cognitive processing of information, which can be subdivided into attention, categorization, memory, and learning, and finally (3) action, i.e., the ability to influence the physical world. These systems will have different levels of complexity, varying from simple control loops to extremely complex and nonlinear systems. To achieve this vision we cannot limit ourselves solely to classical control systems, i.e., sensors, microcontrollers, and actuators. Data collection and feedback will have to become much more complex dynamical processes as has been discussed elsewhere, e.g., in the context of sensor networks [1, 2].

Nanotechnologies for Future Mobile Devices, eds. T. Ryhänen, M. A. Uusitalo, O. Ikkala, and A. Kärkkäinen. Published by Cambridge University Press. © Cambridge University Press 2010.

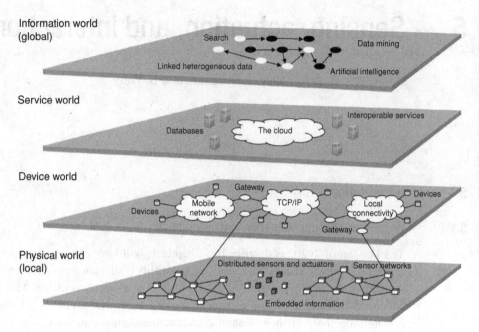

Figure 5.1 Complexity of sensor applications and services in the Internet and mobile communication world.

Figure 5.1 represents the complexity of the architectures of future sensing, processing, and actuation. Primarily, multiple sensors and actuators are connected to local physical and chemical processes. These local sensors will probably be arranged into smart networks with some embedded information processing and memory. Local human interaction with such an environment can include direct responses augmented by the actuation mechanisms distributed in the environment. However, local interaction will become richer when it is connected via some kind of gateway to device networks with more information processing and communication power. These device networks could be based on the Internet, mobile communication, or some local connectivity protocols. As discussed more thoroughly in Chapter 6, networks are merging with seamless connectivity based on cognitive radio. The device network has the dual function of providing a user interface with which to interact with the local physical world and linking the local physical world to the Internet and its services.

There currently exist several sensor applications in different devices, such as game controllers, sports gadgets, and mobile phones [3]. However, our vision presents much more significant opportunities by enabling the connection of locally measured information to Internet services that are able to incorporate this local information into structured global information. Examples of benefits include the real-time tracking of the spread of a disease or epidemic, and interpretation of changes in traffic patterns on roads through the combination of local sensors and the Internet. The Internet is also becoming a massive store of heterogeneous data and information that is increasingly linked in various intelligent ways. Extremely efficient search and data mining technologies are creating a dynamic and real-time map of the physical world and the various economical and social networks.

5.1.2   The need for sensors and actuators in future mobile devices

Sensors can already be found as key features of various battery powered, handheld devices. Location, motion, and gesture recognition are new pervasive elements of applications, user interfaces, and services. One of the enablers of this rapid development has been microelectromechanical systems (MEMS) based on micromachining of silicon (see the review in [3]). The need for low-cost reliable sensors for automotive applications initiated the mass manufacture of silicon MEMS sensors. To date, inkjet printer nozzles, hard disk read heads and micromirror arrays for projection displays based on MEMS components have been the three key applications for actuators on the microscale. The requirements of consumer electronics, especially of mobile phones and game controllers, have driven the miniaturization of MEMS devices further. Today MEMS technologies provide a solid basis for the large-scale deployment of sensor applications. In addition to MEMS, CMOS technologies have also enabled various sensors and, above all, signal processing in these miniaturized devices. The smart sensor system [4] of today consists of transducers, signal conditioning electronics, analog–digital conversion, digital signal processing, memory, and digital communication capabilities (see Figure 5.3).

Industrial automation and robotics together with automotive and aerospace applications have driven the development of actuators and related control systems. There are various interesting materials and mechanisms that are used to create mechanical actuation (see the review in [5]). The choice of mechanical actuator depends on the application, e.g., the required strain and displacement, force, energy efficiency, frequency response, and the resolution in controlling the strain and displacement. In general, an empirical scaling law shows that the faster the actuator, the smaller its strain–stress product seems to be. Finally, the nature of the energy, be it thermal, electrical, magnetic, or photonic, used to control the actuator influences the selection. Mechanical actuators have not yet played a significant role in mobile devices. However, we believe that mechanical actuation will be essential in the development of future user interface concepts. Mobile devices will be able to connect to, and control, several applications of smart spaces where actuation within the physical world is key. Miniaturized devices for health care applications, e.g., insulin pumps and other drug delivery systems, and home automation are good examples of the potential uses of mobile devices to alter the interaction between the human body and its environment.

Nanotechnologies may not completely revolutionize sensor technologies and applications. Existing sensor technologies based on MEMS and CMOS platforms have not yet fully reached their potential to provide sensor applications and networks that improve the human everyday environment. However, nanotechnologies, i.e., different nanoscale building blocks and fabrication processes, will affect the development of sensors, their signal processing and actuators. Nanotechnologies will extend the applications of sensors to new fields, such as smart spaces, body area networks, remote health care, and pervasive environmental monitoring.

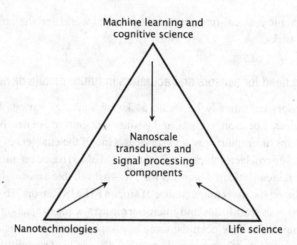

Figure 5.2 Future sensors will benefit from the fast development of nanotechnologies, life sciences, and cognitive science.

In this chapter our goal is to understand and anticipate the impact of current and emerging nanotechnologies on sensors, signal processing, and actuation. This will draw on insights emerging from the life sciences, cognitive sciences, and machine learning. Figure 5.2 summarizes our vision that sensors of the future will use nanoscale electronics, mimic biological systems, and implement signal processing based on robust adaptive algorithms embedded in their physical structures. In the search for these new solutions, we will need to provide answers to the following questions:

- Can we improve existing transduction principles (accuracy, frequency response, etc.) at the nanoscale? Can we find new transducer principles by using nanoscale systems? Above all, can we create sensors that do not exist today?
- Can we integrate transducers in a new way? Can we create complex multisensor arrays using nanoscale building blocks? Can we merge sensors with device surfaces and structural materials?
- Will it be possible to create sensors and sensor arrays with intrinsic capabilities to process and classify their signals into more meaningful information? Can we create sensors with intrinsic energy sources? What are the optimal solutions to integrate energy sources, transducers, signal processing, and communication technologies into miniature systems?
- Can we develop a successful macroscopic actuator from nanoscale building blocks?
- Can we learn something relevant from nature's principles of sensing, signal processing, and actuation? Can we efficiently mimic nature when creating artificial sensors? Which natural processes are efficient and where can we find improvements using optimized artificial systems?
- Can we integrate sensors, signal processing, and actuators into future artifacts and smart spaces in a more affordable and sustainable way using nanotechnologies?

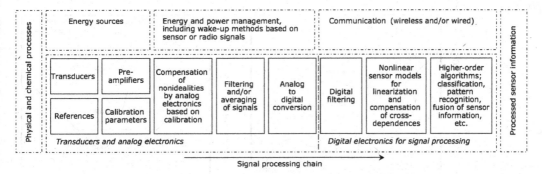

Figure 5.3 Functional partitioning of an integrated, autonomous smart sensor system with sensing, computing, and communication capabilities.

5.2 Sensing

5.2.1 Creating the sensors of the future

Future environments and future devices will have capabilities to sense and analyze their measured data and make decisions based on the analysis. These capabilities will be based on sensors of varying levels of complexity. Meaningful sensors are always connected to a system with signal and data processing capabilities; however, the sensors can vary from simple transducers to smart sensors with computing and communication capabilities. Figure 5.3 shows the functional partitioning of a typical smart sensor system. (For a broader review see [4].) In addition to the sensing and signal processing capabilities, the deployment of sensors in future smart environments requires that the sensors are autonomous in terms of communication and energy management. The integration of all or most of these functions into one miniaturized component – either a single silicon chip or a multichip module – is a clear next phase in delivering integrated sensor technologies to be used in future devices and environments.

Figure 5.3 illustrates the complexity of functions needed for signal processing and interpretation. Transducers are not ideal in their characteristics: they need to be calibrated, different cross-dependences need to be compensated and nearly always the output needs to be linearized. Stable reference materials are needed to support the measurement of quantitative or absolute values. Furthermore, noise from the transducers and measurement electronics requires compensation using modulation, signal averaging, and filtering. Figure 5.3 also shows that the functionality can be partitioned into analog and digital electronics. The trend has been to simplify the analog electronics and compensate for nonidealities using digital algorithms.

The integration of sensors and transducers in devices and other physical structures can also be based on functionality embedded in the structures of the materials. (See Chapter 3 for a review.) In this approach, the sensors become distributed throughout the device structures. An interesting example of a distributed sensor system is the artificial skin concept based on stretchable electronics [6]. Sensors for measuring pressure, temperature, and other parameters can be embedded and distributed in an elastomeric

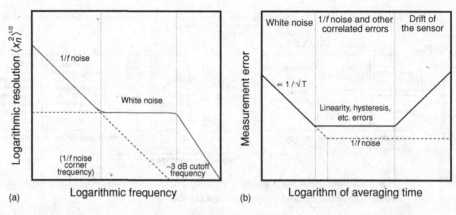

Figure 5.4 (a) Typical resolution (signal-to-noise ratio = 1) of a sensor. At low frequencies the resolution is limited by $1/f$ noise [7, 8, 9]. (b) Effect of averaging on the measurement accuracy of the sensor. Brownian thermal noise can be reduced by averaging, but low-frequency $1/f$ noise and the errors due to nonlinearity and hysteresis are fundamental limits for the measurement accuracy. Finally, the drift of the transducer sensitivity can decrease the measurement accuracy if the underlying phenomena are not understood and compensated.

material that forms the artificial skin of a robot or a human prosthesis. It is envisaged that many chemical or biochemical sensors that have several components with specific sensitivity can also be created in this way. Nanotechnological research is expected to enable very-large-scale parallel transducers and signal processing in miniature systems.

In Section 5.3 we will discuss further this kind of approach where a mixed array of transducers and computational elements for intrinsic signal processing form a network. Typically, such systems are a form of neural network. Before going deeper into the physics of nanoscale sensors and signal processing we will discuss a set of generic figures-of-merit for sensors and sensor systems. First of all, we need to recognize that ultimately it is the application of the sensor that determines the required performance figures. For example, in some applications resolution is more important than the long-term stability of the sensor and vice versa.

The total measurement accuracy of a sensor is a combination of several components: noise, drift, long-term stability, nonlinearity, hysteresis, different cross-dependences (such as temperature dependence), and the calibration uncertainty. All sensors are noisy. The output value of the sensor fluctuates due to several noise sources: fluctuations in the system under measurement, intrinsic fluctuations of the sensor, noise of the measurement electronics, i.e., the preamplifier noise, and the possible noise related to feedback signals. Figure 5.4 summarizes the different characteristics of the overall measurement accuracy of a sensor as a function of signal averaging time. The control of the nonidealities of the transducer characteristics is fundamental for any practical commercial sensor.

In this chapter, we will see that most nanoscale sensors are still in a very early phase of development and commercial applications will have further requirements for sensor technologies. In general, the development of sensors has been characterized by essentially three main problems [3]: (1) how to miniaturize sensors, (2) how to simplify and integrate their readout electronics, and (3) how to package the sensors in a flexible way for any particular applications that may have demanding environmental constraints. The applications set the requirements for power consumption and reliability in addition to the overall accuracy. Chemical and biochemical sensors face particular challenges in reversibility and selectivity as well as in their overall lifetime. Furthermore, commercial constraints, such as production volumes, manufacturability, unit and integration cost, influence their development.

The impact of nanotechnologies in sensor development is thus a very complex problem to analyze. Here we will base our discussion on the possible impact of nanotechnologies on the performance and promise of novel transducer and signal processing principles, however, keeping in mind these challenges of manufacturing and commercialization.

5.2.2   Nature's way of sensing

When considering the design and applications of nanotechnology to sensor development it is useful to look at the sensor/actuator systems found in nature. The design and the complexity of these systems tend to increase with the complexity of the organism but in general there are a number of conserved features. There is often a mechanism to avoid signal-to-noise issues in the form of a threshold effect where a number of stimuli are needed which eventually combine to overcome an energy barrier that then enables transmission of the signal. The noise is also used as an enabler for detection of subthreshold signals as discussed in Section 5.3. Other conserved features include multiple feedback systems between the sensor and the actuator and a hierarchical level of control where many decisions are taken at the same level of the system as the sensory input so that not every decision has to be routed up through the control system to the most complex level (in humans this is the self-aware cognitive level of the brain). We will illustrate these points by considering different sensory systems in nature.

The "simplest" system is to consider a unicellular organism in terms of an intelligent system that performs autonomous functions in relation to changes in its environment. Such systems include planktonic bacteria and unicellular eukaryotes such as yeasts and amoebae. Such isolated cells have a finite number of sensory and response mechanisms and demonstrate intelligence by using a combination of input and feedback mechanisms to demonstrate decision making in response to external stimuli. These decision making processes also include a memory system whereby the learnt responses such as positive reinforcement or negative feedback can be passed from one generation to the next. In addition, there is also a release of signal molecules by the cells to communicate with other individuals from the same species such as the quorum sensing molecules of bacteria.

Bacterial chemotaxis is a well-studied sensory/actuator system whereby a bacterial cell senses a chemical gradient and then responds by swimming either towards or away from the source of the chemicals. Chemotaxis has been used as a model system for the analysis of a number of biological sensing and response systems and has been reviewed in detail [12, 15, 16]. As all of the chemotaxis components function on the nanoscale,

(a)

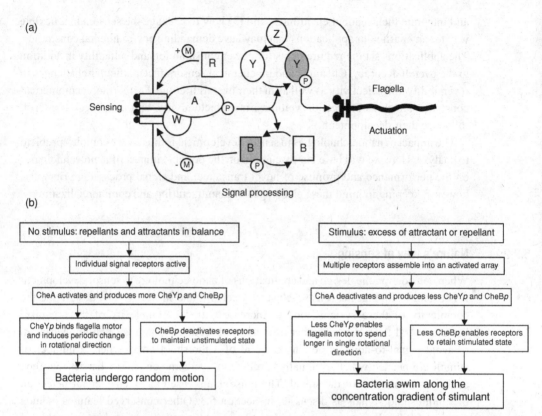

(b)

| No stimulus: repellants and attractants in balance |
| Individual signal receptors active |
| CheA activates and produces more CheYp and CheBp |

| CheYp binds flagella motor and induces periodic change in rotational direction | CheBp deactivates receptors to maintain unstimulated state |

| Bacteria undergo random motion |

| Stimulus: excess of attractant or repellant |
| Multiple receptors assemble into an activated array |
| CheA deactivates and produces less CheYp and CheBp |

| Less CheYp enables flagella motor to spend longer in single rotational direction | Less CheBp enables receptors to retain stimulated state |

| Bacteria swim along the concentration gradient of stimulant |

Figure 5.5 Diagram to demonstrate the sensing, signal processing, and actuation processes of a 'simple' natural system. (a) Diagram of the sensing, signal processing, and actuation components of bacterial chemotaxis. The primary signal processing proteins are CheA (A), CheW (W), CheR (R), CheZ (Z), and the two central regulators CheY (Y) and CheB (B). The phosphate moieties are shown as the letter P in a circle and the methyl moieties are shown as the letter M in a circle. The arrows denote interactions between the components that support signal processing [15, 16]. (b) Flow charts outlining the central steps in the chemotaxis process that allow bacteria to respond to either the lack of chemical stimulus or the presence of attractants or repellants in their environment.

there are obvious insights to be gained on how nanotechnology can be applied to sensing (see Figure 5.5).

The initial sensing events are specific chemical molecules being recognized and binding to specific proteins (receptors) on the surface of the cell. This binding event is converted into a chemical sensing system using a common biological strategy, allosteric control. Allosteric control is the process whereby a signal modification (effector) molecule binds to a site on the protein other than the active site (in the case of the receptor protein, the site where the chemical binds) and this induces a change in the conformation of the protein that modifies its activity (for instance, in the case of an enzyme this may change the rate of catalysis). In chemotaxis the receptors follow the concerted allosteric control model [14] to give an array of active receptors on the inner surface of the bacteria, the size of the array being proportional to the number of receptors that are bound to their signal molecule. This relationship between the number of active receptors and the size of the receptor array on the inner surface of the cell enables the signal processing system to respond in proportion to the strength of the signal, in effect to quantitate the signal. This array creates a microenvironment on the inner surface of the cell specifically designed to activate the signaling pathway by recruiting the signal processing components to one location and enabling the rapid exchange of covalent modifications (phosphate and methyl groups) between the protein components. This chemical modification is another example of allosteric control where the covalent linkage of the chemical group changes the surface chemistry and conformation of the protein, either activating or repressing its function depending on the nature of the modification.

The chemotaxis control system includes both the exchange of phosphate groups between proteins mediated by the CheA phosphorylase enzyme and the exchange of methyl groups mediated by the CheB and CheR methylases. These two processes have different effects on the target proteins and illustrate another common biological control strategy. The signaling components often exist in a dynamic equilibrium of different functional states (varying from inactive to active but with gradations of activity in between) and movement between these states correlates with the addition of chemical modifications and other effector molecules that move the proteins through successive conformations. One type of modification increases the protein activity (in the case of chemotaxis addition of phosphates to either CheB or CheY increases their activity) while the other decreases the activity (the methylase activities of CheR and phosphorylated CheB modify the response of the receptors to the chemical signal). The two pathways, one stimulatory and one inhibitory, maintain the system in a dynamic equilibrium. When an external signal is detected this perturbs the equilibrium state so that one pathway then increases in turnover (e.g., increased phoshorylation of CheY) while the other decreases and this change in the balance between the two pathways acts to amplify the signal, increasing the rate of response to the external signal. Fine tuning this equilibrium state enables complex control behaviors. For instance the chemotaxis system uses this equilibrium to achieve four different control processes: adaptation, sensitivity, gain, and robustness [16].

The final step of the process is the link between signal processing and actuation. Here chemotaxis shows another biological control strategy, compartmentalization. The sensor array is physically separate from the flagellar motor complexes in the cell and the activated messenger protein, phosphorylated CheY (CheYp), has to diffuse between the two sites to change the behavior of the flagella actuator. This movement of components between different compartments within a biological cell is an extremely common control mechanism and the use of compartments allows the cell to create different coordinated chemical microenvironments that support very different catalytic reactions which take place in parallel.

The control mechanisms shown in Figure 5.5 all enable the individual cell to respond to a change in the chemical environment in a time frame of seconds to minutes. There is a further level of control over the system driven by the use of the cell's memory system, the DNA genome. Each protein component is encoded by the genome and this code is converted into a protein by the processes of transcription and translation. For example, in the chemotaxis system, transcriptional control allows the cell to change the distribution of the types of receptor proteins expressed on the surface and thus the identities and the range of concentrations of the signal molecules to which the system will respond. Transcription and, to a different extent, translation are controlled by similar equilibria and feedback loops and their effects are seen over a longer time frame of minutes to hours. In addition genetic rearrangements can change the availability of components (seen as genetic mutation), which changes the behavior of the chemotaxis system from generation to generation of bacteria. All of these levels of control give the biological system exquisite ability both to control their response to environmental changes in a short time frame and also to adapt to changes over the longer term, and to carry these adaptations from one generation to the next. Nanotechnology could be used to apply these biological control strategies to artificial devices, but such an approach is still in its infancy.

Bacterial chemotaxis is a behavior of individual cells. Many of the control strategies are found in multicellular organisms but in this case these different functions have often become distributed over multiple cell types, the number depending on the organism and the nature of the sensory systems. Some of these cells show very strong specialization with the development of particular cellular structures and morphologies and specific patterns of gene expression to give the functionality, specificity, and signal amplification necessary to demonstrate a sensing/actuation response at the scale of a multicellular organism. The other feature of multicellular organisms is the emergence of a nervous system and the increasingly complex control mechanisms that a nervous system enables. This can be demonstrated by consideration of the mammalian taste or olfaction systems [10, 11].

Similarly biological actuators show a high level of diversity. At the cellular level a common response is movement where the motion is often driven by use of molecular motors such as flagella but also by rearrangements of the cellular cytoskeleton. The amplification of molecular movement at the nanometer scale to detectable motion and measurable force on a millimeter to meter scale is a triumph of biological design and provides a benchmark for the development of nanotechnological actuator systems.

Natural systems are the results of evolutionary processes that have given increasing complexity. However, the results of evolution do not necessarily represent the optimal solution in all aspects. Nor are they fixed, the process of natural selection shows a consistent level of noise due to mutation from cell to cell and generation to generation. This rate of mutation results in any one generation showing a number of phenotypes and these behave differently depending on the environments encountered. If a particular set of phenotypes confers advantages on the individuals expressing them in a particular environment, then these will be rapidly selected and the phenotypes will expand across the population, while they continue to confer an advantage in that environment. Overall natural selection maintains a number of different functions across the population of individuals and presents a rapid system for selecting useful functions. This mechanism does not yet have a parallel in nanotechnological applications.

Biological structures are typically modular, using nearly similar building blocks in large numbers to generate complex functional molecules, cells, organs, and whole organisms. These building blocks typically are used in and form hierarchies. Despite this, the functionalities of the biological systems are not modular: very complex regulatory networks exist at various scales of interaction. Stochastic processes play an essential role: Brownian noise and fluctuations provide the mechanisms for changes in the molecular level functionality and it is possible to use thermodynamics to model these biological systems with free energy as the proper quantity. We can immediately see some similarities with proposed nanotechnological approaches: large and complex arrays of nearly similar elements together with complex interaction mechanisms and stochastic phenomena within these arrays. The designers of architectures to create robust and fault tolerant sensing and computing solutions can potentially learn from natural systems. Natural sensory systems are extremely efficient at some tasks, such as recognition of complex patterns and associative memory. In many of these cases the efficiency is related to parallel processes and optimized hierarchy of information processing. The sensitivity of natural systems in detecting physical or chemical stimuli can be also very high and very challenging for artificial sensors to match. Chemical sensory systems in nature are also very robust and self-healing. However, it is worth remembering that natural systems have their limitations and are in a state of evolutionary flux in an effort to respond to the environmental changes that determine their performance criteria. Biology can provide inspiration and a vision of the level of complexity and efficiency that can be achieved but it should not confine human imagination; additional solutions to sensory problems await discovery outside the natural world.

5.2.3   Physical nanoscale transducers – nanowires and resonators

There is no doubt that mechanical transducers will play a key role in building new user interfaces and interaction paradigms for future electronics and ambient intelligence. Detection of movement and touch have already become an essential part of user interfaces. Accelerometers, angular rate sensors, pressure sensors, and strain sensors are manufactured in large volumes based on silicon micromachining and other low-cost manufacturing solutions. The dimensions of micromechanical accelerometers and pressure sensors, including the signal conditioning electronics, are in the order of 1 mm^3. This level of miniaturization is sufficient for most electronic devices, such as mobile phones and mobile multimedia computers. What is the additional benefit of further miniaturization of transducers and what are the practical and theoretical limits? How far does the miniaturization make sense? Are there emerging drivers for developing new mechanical transducers?

A mechanical transducer is a macroscopic mechanical element that converts force into an electrically or optically measurable change [13]. For example, a system of a proof mass and a spring converts acceleration into a deflection in the spring. The deflection can be detected by the strain in the spring or the displacement of the proof mass. In a similar manner, a deflecting thin diaphragm can be used for measuring a pressure difference across it. Commercially available state-of-the-art accelerometers and pressure sensors use either capacitive displacement measurement or a piezoresistive transducer to convert the mechanical force into electrical voltage or current.

Nanoscale electrical and electromechanical structures are attractive for nanosensors, promising reduction of device size, improved performance, and lower power consumption. In addition, such nanoscale structures can possibly be arranged into arrays of similar components together with signal processing, signal amplification, and memory elements made of the same or compatible materials. These arrays of nanoscale devices may form a new foundation for fault-tolerant, robust, energy-efficient sensing and computing. Functional materials will be another way to integrate sensing function into mechanical structures of macroscopic devices. Nanoscale mechanical structures have the advantages that they allow: (1) interaction with molecular level structures; and (2) exploitation of unique nanoscale physical properties to interact with the macroscopic environment.

We will review some of the most interesting sensor elements: nanoscale strain gauges, nanowire field effect transistors (FETs), nanomechanical resonators, and carbon nanostructures (carbon nanotubes (CNTs) and two-dimensional graphene). It is very important to notice that these nanoelectromechanical systems (NEMS) are not merely downscaled MEMS devices. NEMS represents a disruptive solution based on new fabrication technologies, new physical effects, and new system concepts that become possible at the nanoscale. There are two clear strategies for realizing nanoscale electromechanical devices and sensors: (1) top-down processing of nanoscale structures, i.e., MEMS miniaturization using new fabrication principles, and (2) bottom-up growth of meaningful structures for nanoelectromechanical devices.

5.2.3.1   Nanoscale strain gauges based on carbon nanostructures

In general, the miniaturization of accelerometers, angular rate sensors, and pressure sensors is ultimately limited by the signal-to-noise ratio (SNR) which scales inversely proportionally to the size; e.g., a larger inertial mass improves the SNR for the spring-based test mass. Furthermore, the continued miniaturization of MEMS transducers is not feasible; capacitive transducers are sensitivity-limited due to the reduced area and piezoresistive transducers are limited by increasing noise due to increasing resistance. In order to miniaturize mechanical sensors it is essential to find new and more efficient transducers.

CNTs and especially single-walled CNTs (SWCNTs) are interesting candidates for sensor applications owing to their unique electrical and mechanical characteristics. CNTs are typically 1–50 nm in diameter (SWCNTs 1–3 nm) and their length can be up to centimeters, though a few micrometers is more typical. Their Young's modulus is extremely high ~ 1.2 TPa, their tensile strength is ~ 45 GPa and their elastic maximum strain ϵ_{max} is $\sim 6\%$. SWCNTs can demonstrate both metallic and semiconductor properties with an electron mean free path of $\sim 1\,\mu$m and can support electric current densities up to $\sim 10^9$ A/cm^2. The use of CNT-based piezoelectric transducers in mechanical sensors has been demonstrated with good results [24, 27, 28, 29]. Reversible electromechanical characteristics of CNTs can demonstrated, e.g., by manipulating a nanotube fixed at both ends to metal electrodes using an atomic force microscope (AFM) tip.

The electrical resistance of SWCNTs is very sensitive to the strain ϵ. The dependence can be modeled theoretically based on the distortion of the atomic structure of the metallic SWCNT which affects the bandgap E_{gap} by $dE_{\mathrm{gap}}/d\epsilon = \tilde{E}_{\mathrm{gap}} \approx 80-100$ meV/%. The

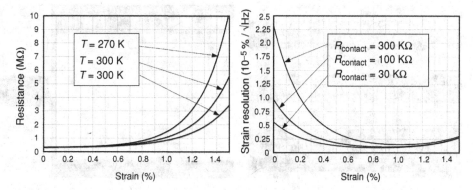

Figure 5.6 (a) Resistance of a SWCNT at three different temperatures based on (5.1). Here $R_{\text{contact}} = 300\,\text{k}\Omega$. (b) Calculated strain resolution $\langle \epsilon_n^2 \rangle^{1/2} \approx (\partial R / \partial \epsilon)^{-1} \cdot (\partial V / \partial R)^{-1} \cdot \sqrt{4k_{\text{B}} T R(\epsilon)}$ of the metallic SWCNT in a bridge measurement configuration ($R_{\text{bridge}} = 500\,\text{k}\Omega$) with 50 mV bias and optimized impedance matching to the preamplifier input impedance. In these calculations the contact resistances range from 30 kΩ to 300 kΩ and $T = 300$ K.

resulting change in resistance can be two orders of magnitude for a strain in the range of 1–3%, according to the thermal activation model [25, 26]

$$R(\epsilon) = R_{\text{contact}} + \frac{h}{8|t^2|e^2} \left[1 + \exp\left(\frac{\tilde{E}_{\text{gap}}\epsilon}{k_{\text{B}} T} \right) \right], \tag{5.1}$$

where $|t^2| \approx 0.2$ is the transmission coefficient, $h/(8|t^2|e^2) \approx 16\,\text{k}\Omega$ and R_{contact} is the contact resistance. Typically, the contact resistance is very large $R_{\text{contact}} \sim 100\text{–}400\,\text{k}\Omega$ and is a challenge for device development. Figure 5.6(a) illustrates the nonlinearity and the approximate temperature dependence of the strain–resistance characteristics of a metallic SWCNT. Due to the increasing resistance and thermal noise $4k_{\text{B}} T R(\epsilon)$, an estimate optimum of strain resolution is achieved at some finite value of strain, as shown in Figure 5.6(b).

In practice, the use of SWCNT strain transducers is limited by four issues: (1) the contact resistances have so far been very high resulting in thermal noise; (2) the nonlinearity of the strain–resistance curve of (5.1) limits the sensitivity at small values of strain ϵ; (3) the temperature dependence is large based on (5.1); and (4) the low-frequency excess noise (1/f noise) has been very high in all the SWCNT noise measurements [30]. The integration of the SWCNT transducer into a micromechanical structure determines the displacement resolution of the sensor, $\langle x_n^2 \rangle^{1/2} \approx (\partial \epsilon / \partial x)^{-1} \cdot \langle \epsilon_n^2 \rangle^{1/2}$.

Despite this, SWCNTs are likely to continue to be evaluated as components for strain gauges in various MEMS and macroscopic devices. To overcome the low-frequency noise and temperature dependences, more complex measurement schemes are needed, such as alternating current bridges to compensate for low-frequency noise and temperature dependences and even to force feedback to linearize the sensor output. With these, SWCNT transducers may become enablers for the next level of miniaturization of MEMS accelerometers, pressure sensors, and cantilever biochemical sensors.

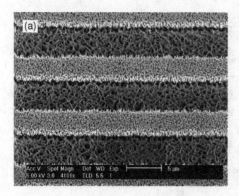 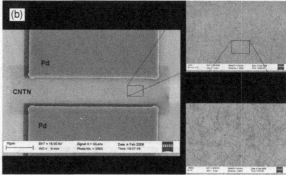

Figure 5.7 (a) Patterned ZnO nanowire arrays for tactile sensing (Nokia Research Center and University of Cambridge). (b) Carbon nanotube network (CNTN) based transistors and circuitry demonstrated on a flexible substrate (Nokia Research Center and Helsinki University of Technology).

5.2.3.2   Piezoelectric nanostructures

Mechanical–electrical triggers, sensors, and energy harvesting devices have been demonstrated [31, 32, 33, 34, 35, 36, 37, 46] based on nanoscale piezoelectric nanowires and nanobelts of zinc oxide (ZnO), gallium nitrite (GaN), or perovskites, such as barium titanate ($BaTiO_3$). ZnO and GaN have relatively low piezoelectric constants, ~ 12 pC/N and ~ 3 pC/N, respectively. The longitudinal piezoelectric constants of perovskites are much larger: single-crystal $BaTiO_3$ has a charge constant ~ 85 pC/N though this is still short of the widely used piezoelectric compound lead zirconate titanate (PZT, ~ 268 pC/N).

Piezoelectric nanostructures can be used either as mechanical transducers directly converting force into electrical charge, or as NEMS resonators actuated by an external electrical field. A piezoelectric element can be modeled as a current source i in parallel with a capacitance. The current source $i = d(\partial F/\partial t) = dAE(\partial \epsilon/\partial t)$, where d is the piezoelectric constant, F is the applied force, A is the load area, E is Young's modulus, and ϵ is the strain. Thus the resulting signal is proportional to the rate of the applied force or strain. The key challenge in creating practical sensing and energy harvesting circuits based on ZnO nanowires is to solve how to detect and exploit the tiny currents created by the mechanical deformations.

Figure 5.7(a) shows an experimental array of vertical ZnO nanowires that are patterned to form conducting stripes. The crossbar structure of the patterned ZnO nanowire stripes can be used to form a low-cost and environmentally friendly generic technology to develop touch-sensitive surfaces that are optically transparent. Such an array of ZnO nanowires can be deposited on a surface of various substrates (silicon, glass, PET) at a relatively low temperature (roughly $70-100\,^{\circ}$C). The technology should support the development of transparent, flexible, touch-sensitive active matrix arrays on top of flexible polymer displays or other flexible polymer substrates.

One of the key advantages of a piezoelectric touch-sensitive surface is that as the mechanical excitation generates electrical power [31], the energy consumption of such

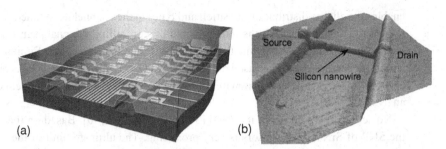

Figure 5.8 (a) Basic structure of a SiNW FET. (b) AFM image of a double-gate SiNW-based FET for experimental biosensors (Nokia Research Center, Cambridge) [41].

a device should be significantly smaller than that of current technologies. ZnO nanowire power generation depends on the amplitude and the frequency of the mechanical excitation. The approximate current generated by a single ZnO nanowire is roughly 0.1 pA at ultrasound frequencies (~ 40 kHz). If the density of the ZnO nanowires in a given area were $\sim 10^9/cm^2$, the maximum available current would be 0.1 mA/cm^2. In practice, the strain amplitudes can be higher and the excitation frequencies much lower. Experiments based on the ZnO devices shown in Figure 5.7(a) show transient current signals in the range of 10–100 nA/mm^2 in the real touch sensor configuration. Both material development and careful engineering are needed to achieve practical devices which can be applied in the real world.

Another possible way to use ZnO nanowires for sensing is based on the change of the transport characteristics when the nanowire is bent [37]. Bending and flexing stresses induce piezoelectric potential across the wire and create additional strain-induced resistance that combine to influence electrical transport. However, ZnO nanowires require relatively high bias voltages (> 1 V) to become conductive. Furthermore, ZnO nanowires can be arranged into transistor configurations with power amplification.

5.2.3.3 Nanowire field effect transistors and sensors

Semiconducting nanowire field effect transistors (NW FETs) are very promising nanoscale components for chemical and biochemical sensors. NW FETs can be used both for detecting events with spatial resolution at molecular level or simply for increasing the charge sensitivity of the chemical sensor. The dimensions of the gate area are in a similar range to those of biological macromolecules, and the surface-to-volume ratio of these structures is very large, making surface effects dominant over the bulk properties.

Figure 5.8(a) illustrates the basic structure of a silicon nanowire (SiNW) FET that is typically 10–100 nm in diameter and covered by a thin (< 10 nm) SiO$_2$ layer. If the SiNW is surrounded by a gate electrode, the conducting channel can be controlled by the gate voltage V_g. The sensing mechanism of a NW FET is then based on the controlled modification of the local electric field in the channel due to the binding of charged molecules to the nanowire surface.

Figure 5.8(b) shows an experimental device [41] for studying SiNW FET characteristics for chemical and biochemical sensors. The device is a double-gated structure that enables the charge distribution (position and type) in the channel to be tuned. Optimization of the device sensitivity is thus possible by controlling the charge in the channel next to the chemically functionalized surface of the channel region. Similar types of structures can be developed based on SWCNT, ZnO, SnO_2, In_2O_3, and GaAs nanowires [44, 45, 46]. Arrays of such nanowires have been used for principal component chemical analysis [47, 48].

Noise in SiNWs has been modeled [39] and measured [40]. Based on these results, the SNR of SiNW FET sensors is very promising. The ultimate limit of the resolution of nanowire sensors is set by the low-frequency $1/f$ noise. Studies [30, 40, 42] have shown that $1/f$ noise in CNTs, metal nanowires, and SiNWs seems to behave according to well-known laws of low-frequency noise [7, 8, 9]: $1/f$ noise seems to be inversely proportional to the contact interface area and to other interfaces with possible defects creating a trapping–detrapping process of charge carriers. Even though the origin of the $1/f$ noise in general is not yet fully understood and more careful noise measurements of nanoscale systems are needed, the known scaling laws seem to be valid.

The use of CNTs and two-dimensional graphene in creating chemical sensors has been studied by several groups [43]. The basic idea is the same: by functionalizing CNT or two-dimensional graphene with chemical groups that can bind charged molecules, the transport characteristics of the CNT or two-dimensional graphene can be changed. The key challenge here is to create transducers with sufficient specificity and stability. Figure 5.7(b) presents a CNTN-based transistor structure that can be used as a basis for the development of chemical and biochemical sensors. The transistor characteristics can be controlled by developing thin monolayer coatings on top of the sensitive area. Experiments in which CNTs are coated with atomic layer deposition have shown that the effect of the chemical environment (e.g., humidity) on the device can be controlled [49].

5.2.3.4   Chemical sensing with nanowire, carbon nanotube, and graphene devices

Since their development there has been considerable effort into using the properties of nanowires for sensor applications [17, 23]. The two leading methods use nanowires in FETs or as nanoresonators. The Lieber research group has investigated the use of silicon oxide nanowires as FETs and the functionalization of these nanowires to enable specific sensing capabilities [19, 38]. They have proved that nanowire FETs can be exploited to measure classes of targets that include ions, small molecules, proteins, nucleic acids, and even viruses with high sensitivity and selectivity [20, 38]. Figure 5.8 shows a representation of a nanowire FET sensor array with different surface functionalized nanowires arranged in microfluidic channels to allow sample access to and binding on an individual nanowire.

One of the reasons why a wide range of targets has been reported is that the silicon oxide nanowires can be chemically functionalized using a range of techniques that had previously been developed and tested at the macroscopic scale on planar glass and silicon oxide surfaces. The selection of the correct surface functionalization is then driven by ensuring that any change in charge at the surface of the FET nanowire is due to a specific binding event. This places some limitations on the design of the sensor. For example, the sensitivity of the silicon oxide nanowire FET is significantly lowered at the ionic strength of most biological fluids. This can be addressed by appropriate sample preparation before the addition of the sample to the nanowires, but this extends the time required for the analysis and introduces potential sources of error. In response a number of strategies are being developed to overcome this limitation [20].

The success of nanoresonator devices relies on the functionalization approach not changing the mechanical properties of the nanowires so that the sensor output can still be compared to a control signal within the context of the device [18]. Nanoresonators have been made from a number of different materials and each has different functionalization chemistries available for use. As with the FETs, the most common approach is to use functionalization first developed at the macroscale and apply it to the nanoscale. A notable exception are resonators based on CNTs, or graphene where a number of new chemistries have been developed [22]. A number of published nanoresonator studies do not utilize a functionalization strategy but instead rely on direct physisorption of the analyte to the resonator. This avoids biases from the functionalization but raises the problem of consistency between different batches of resonators.

Nanowires have been applied to sensing for almost 10 years but very few nanowire sensors have made it to the market. One barrier is the complexity of manufacturing the resonator arrays as, though a number of methods have been developed for aligning large numbers of nanoresonators on a surface, none has yet been developed into a manufacturing process [20]. The full potential of nanoresonator sensors will rely on the development of novel device architectures that account both for the envisaged multiple functions of the nanoresonators and enable sample processing and access where a section of the architecture is being used as a sensor. The small scale of an individual nanoresonator or NW FET opens up the possibility that they can be used to measure different analytes in parallel, known as multiplexed measurements. This would require a range of functionalization chemistries to be applied in the manufacturing process and the device optimized to allow for the differences in sensor performance that this would introduce at different places in the architecture. These represent significant challenges. These issues have been approached from a phenomenological perspective but a theoretical framework is starting to emerge that attempts to understand and predict the behavior of nanoscale sensors [21].

The best nanowire functionalization strategy for a particular sensor will be a compromise influenced by the identity of the analyte, the chemistry of the nanowire, the analytical procedure to which the sensor contributes, the manufacturing process, and the quality control requirements. Functionalization represents a considerable challenge to the mass production of sensors based on NW FETs and nanoresonators.

5.2.3.5 Mass spectrometry using nanoscale resonators

Nanoscale electromechanical resonators are perhaps the most promising components for creating high-resolution sensors. An electromechanical resonator is a mass–spring system that resonates with a characteristic frequency $\omega_0 \propto \sqrt{\kappa_{\text{eff}}/M_{\text{eff}}}$, where M_{eff} is the effective vibratory mass and κ_{eff} the effective spring constant. The actuation force is typically capacitive, magnetic, or piezoelectric. Development of the (nano)electromechanical resonators has been focused on the fabrication of devices with higher resonance frequency (towards 1 GHz) with high Q values and on the development of practical oscillators with low phase noise and low temperature dependence. Advances in MEMS reference oscillators in the 10–60 MHz range are very promising [50, 51, 52, 53].

Nanoscale resonating devices have been developed [54, 56, 57] based on nanowires of different kinds (silicon, metal, metal oxide) and carbon nanostructures. Thus far the devices have been studied in the context of the exploration of physical principles. However, there are experiments that demonstrate the use of the nanowire resonators as ultrasensitive mass sensors. Resonating nanowire mass sensing has some advantages: the resonating structures are smaller and thus the relative change in the mass δm of the resonator is larger; the resonance frequency is higher, improving the measurement resolution, and possible new readout schemes can be developed, based on nanoscale electronics.

The principle of operation of resonating mass sensors is very simple. The additional mass of the resonating structure δm is proportional to the resonance frequency change $\delta m = (\partial \omega_0 / \partial M_{\text{eff}})^{-1} \cdot \delta \omega \approx (2 M_{\text{eff}} / \omega_0) \cdot \delta \omega = \sqrt{4 M_{\text{eff}}^3 / \kappa_{\text{eff}}} \cdot \delta \omega$. Smaller effective masses M_{eff} and larger resonance frequencies ω_0 improve the mass sensitivity of the resonating mass sensor.

However, the mass sensitivity $\partial \omega_0 / \partial M_{\text{eff}}$ is not a sufficient figure of merit for evaluating the performance; the mass resolution of the resonator is needed – including the different noise sources discussed in [55, 56, 57]. The external mechanisms for energy dissipation are related to gas damping, energy leakage to supports, and transducer coupling losses. Intrinsic losses are related to the mechanical dissipation (phonon–phonon, phonon–electron interactions) present in any perfect crystalline material enhanced by bulk and surface defects. Figure 5.9(a) presents an empirical law of the quality factor Q of monocrystalline mechanical resonators with size ranging from macroscopic to the nanoscale. The Q factor is defined as 2π times the ratio of the time-averaged energy stored in the resonator to the energy loss per cycle. The Q factor decreases roughly with increase of the ratio of volume to surface area, and it is known that $Q \propto M_{\text{eff}} \cdot \omega_0 / A_{\text{eff}}$, where A_{eff} is the effective area perpendicular to the motion. Smaller resonators are also more easily affected by acoustical energy leakage to the anchors of the vibrating structure.

An estimate of mass resolution can be calculated by taking into account the power spectral densities of the frequency fluctuations of the resonator due to the intrinsic and preamplifier noise, $S_\omega(\omega) \approx S_{\omega,\text{int}}(\omega) + S_{\omega,\text{preamp}}(\omega)$. Thus we can write the mass resolution as

$$\langle m_n^2 \rangle^{1/2} \cdot \sqrt{\Delta f} = \left| \frac{\partial M_{\text{eff}}}{\partial \omega_0} \right| \cdot S_\omega(\omega)^{1/2} \cdot \sqrt{\Delta f} \approx \frac{0.8 M_{\text{eff}}}{Q \cdot x_{\text{c}}} \sqrt{\frac{k_{\text{B}} T}{M_{\text{eff}} \omega_0^2} + \frac{\omega_0 S_V(\omega)}{Q (\partial V / \partial x)^2}},$$

(5.2)

where x_{c} is the maximum linear drive amplitude [53, 57], $S_V(\omega)$ is the power spectral density of the preamplifier voltage noise, $\partial V / \partial x$ is the displacement-to-voltage transfer function, and $\Delta f \approx f_0 / Q$ is the bandwidth.

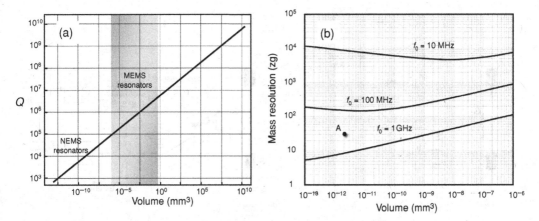

Figure 5.9 (a) Empirical law of measured quality factors (Q factors) of several monocrystalline mechanical resonators as a function of resonating element volume. The Q factor decreases roughly inversely proportionally to the ratio of volume to surface area. Adapted from [57]. (b) Mass resolution $\langle m_n^2 \rangle^{1/2} \cdot \sqrt{\Delta f}$ of the resonating nanowire by (5.2) as a function of the effective volume $V_{\text{eff}} = 0.735 w_{\text{nw}} t_{\text{nw}} l_{\text{nw}}$ for three different fixed resonance frequencies. For small devices the intrinsic noise dominates. For large devices the mass resolution is determined by the amplifier noise contribution. Here $\partial V/\partial x = 0.1$ nV/fm, $\sqrt{S_V(\omega)} = 1.5$ nV/$\sqrt{\text{Hz}}$, $T = 300$ K and $p = 1$ atm. For an example (see Point A), if $M_{\text{eff}} = 5$ fg, $f_0 = 300$ MHz and maximum linear drive amplitude $x_c = 3$ nm, we get $Q \approx 1000$ and the mass resolution $\langle m_n^2 \rangle^{1/2} \cdot \sqrt{\Delta f} \approx 30$ zg.

Figure 5.9(b) presents improvement in mass resolution as a function of the effective volume of the resonating structure. The calculated values are in good agreement with the best measured results for flexural MEMS resonators [60, 63, 67]. Miniaturization down to nanoscale is beneficial because of the smaller M_{eff} but requires simultaneously higher resonance frequencies ω_0.

When designing a practical measurement system, the resonator and the transducer need to be configured very carefully to maximize $\partial V/\partial x$ and x_c, within the available energy budget. A compromise with the stiffness κ_{eff} of the resonating structure may thus need to be made; (5.2) can be used for optimizing the mass sensor system. The optimal resonance frequency depends on the size of the resonator and the coupling to the amplifier, as seen in Figure 5.9(b) and from (5.2). NEMS resonators provide an excellent technology for very high-resolution mass sensors and, while the measurement systems (see Figure 5.10) remain complex, there is progress on incorporating them into analytical instrumentation [58].

The results for high-quality MEMS resonators show that the use of flexural mode resonators has not been the optimal strategy towards very high Q values. This is due to the gas damping and the energy leakage at the anchor points. The use of different geometries, such as bars and disks, where the vibrational modes are longitudinal or otherwise bulk mode vibrations, have been a way to improve Q values [50]. In these geometries the coupling of external electrical excitation has been the challenge. So far all the NEMS resonators have been based on flexural oscillations that are much easier to

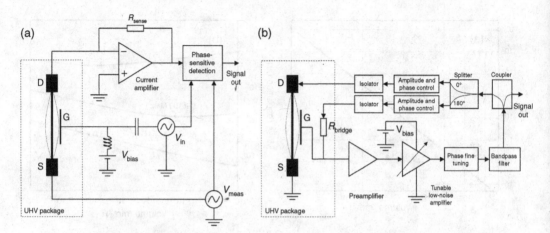

Figure 5.10 Possible measurement systems for a resonating silicon nanowire, enabling ultrahigh resolution: (a) heterodyne measurement based on two external oscillators (according to [60]); (b) self-sustaining UHF NEMS oscillator based on an on-chip bridge resistor R_{bridge} to compensate the resonator intrinsic resistance (according to [61]).

accomplish in the form factors of nanowires or CNTs. An interesting research challenge is to try to find other nanoscale geometries of mechanical oscillators – with improved Q values.

The use of bottom-up grown SiNWs for resonators has been studied [59, 60, 62]. A CNT-based resonating mass sensor has also been developed based on the same principle [63] while two-dimensional graphene or graphene-oxide NEMS resonators [64, 65] have been studied as other possible material systems for high-Q resonating sensors. It is anticipated that nanotechnologies will create new material solutions as well as new nanoscale mechanical structures that will be cost effective for mass production devices.

A self-sustaining oscillator based on silicon nanowire resonators has been demonstrated [61], see Figure 5.10(b). Even though the device requires complex circuitry to tune the closed loop operation, this demonstration of a self-sustained NEMS oscillator simplifies the mass detection electronics and gives the promise of nanoscale implementation of self-sustained oscillators in the future. A fully nanoscale oscillator would be the enabler for chemical and biochemical sensors and for their signal processing.

5.2.3.6   Field emission based detection

A very interesting new transducer principle based on field emission by CNTs has been introduced [66, 67]. The resonating field emission diode (FED) formed by a CNT in an electric field acts at the same time as an antenna, a resonator forming a bandpass filter, an amplifier, and a rectifying diode. The electric charge in the CNT concentrates at its 'sharp' tip when the CNT is placed in an external electric field [68, 69, 70]. The strength of the external electric field ($E_{ext} \sim 10^8$ V/m) determines the probability of the electrons tunneling through the vacuum from the cathode to the anode. The resulting current I depends on the variable gap g between the anode and

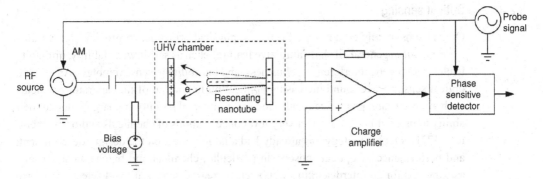

Figure 5.11 Measurement of CNT resonance based on field emission current, according to Jensen *et al.* [66, 67].

cathode, $I(g) = c_1 A(\gamma E_{\text{ext}}(g))^2 \exp(-c_2/\gamma E_{\text{ext}}(g))$, where c_1 and c_2 are constants, A is the emitting area, and γ is the local field enhancement factor.

Figure 5.11 shows the measurement setup that can be used for an ultrahigh-resolution mass sensor [67]. Jensen *et al.* have demonstrated a mass resolution of 0.13 zg$/\sqrt{\text{Hz}} = 0.4$ Au atoms$/\sqrt{\text{Hz}}$ at room temperature. The high resolution is based on the sensitivity of the field emission detection and the low mass of the CNT resonator yielding three orders of magnitude higher mass responsivity compared to SiNW structures, $(\partial \omega_0/\partial M_{\text{eff}}) \approx 1$ MHz/zg. Typical resonance frequencies f_0 are from 100 MHz to 1 GHz depending on the length of the CNT.

The system can be used for miniaturization of mass spectrometers that are currently complex and bulky and require ionization of measured molecules. The measurement principle also allows a possible simplification of the measurement system for applications that do not require quantitative resolution. Such an array of resonating nanotubes with a very simple diode detector (a nanotube FED together with a subsequent integrator) can create a measurement system for chemical/biochemical sensing. Very interesting field emission driven self-oscillating circuitry based on SiC nanowires for dc–ac conversion has been demonstrated [71], although the bias voltage levels are very high (~ 100 V).

5.2.3.7 From nanowire and NEMS devices to complex nanoscale sensor systems

A huge amount of development is still needed to create commercially feasible nanoscale transducers. In theory nanoscale transducers can be integrated with mixed signal microelectronics and different MEMS (microfluidics, diaphragms, inertial masses, and actuators) to build complete sensor microsystems. The design of the packaging of the sensors will also be vital: miniaturized, microscale hermetic or vacuum packaging and microfluidics systems will be essential for these future sensor systems.

In Section 5.3 we will further discuss the creation of complete sensor systems with nanoscale signal processing elements. If these are successful, complex intelligent arrays of sensing, computing, and memory elements can be built not only to sense but also to interpret the sensory information.

5.2.4   Optical sensing

Optical sensors rely on a range of different physical principles to provide signal transduction, but in general all share attractive features such as sensitivity, stability, immunity to electromagnetic interference, robustness, and the ability to sense remotely. As a result, optical sensors have found increasing application in and out of the laboratory in fields such as healthcare, pharmaceuticals, and environmental monitoring, largely due to their ability to discriminate and detect very low concentrations of chemicals and/or biochemicals [72]. Nanotechnology has already had a large impact on the design, development, and performance of optical sensors, in particular, chemical and biochemical sensors, as expected for an interdisciplinary subject. Increased sensitivity and selectivity based on nanotechnological advances have already been demonstrated and this trend is set to continue.

5.2.4.1   Optical chemical sensors

Optical chemical sensors separate into two main classes of operation; those performing direct detection and those that indirectly detect the analyte via its interaction with an intermediary. Direct detection senses an intrinsic property of the analyte, either its complex permittivity (refractive index and/or absorption) or luminescence emission (fluorescence, phosphorescence, or Raman), whilst indirect detection monitors the change in optical properties of the intermediate moiety upon analyte coupling, and hence is useful for analytes with no intrinsic optical response. There is an analogous division for optical biosensors, which employ either labeled or label-free approaches. Label-free detection has the advantages that the analyte is unmodified (so the system is undisturbed) and that quantative, kinetic information can be gathered [73]. In the labeling approach either the analyte or a biorecognition molecule is tagged with a (often fluorescent) marker, whose emission indicates coupling and thus detection. Nanotechnology has had an impact here as the traditional fluorescent dye markers are being supplanted by quantum dots, metallic nanoparticles, and even Raman active CNTs, all of which exhibit superior photostability and can deliver single-molecule sensitivity.

Quantum dots are nanocrystals of semiconducting material, typically of the order of 2–50 nm in diameter, and they have received increasing scientific and technological interest due to their unusual optoelectronic properties. These properties arise from the tight spatial confinement of electron–hole pairs (excitons) within the dot which reduces the excitonic Bohr radius from that of the bulk semiconductor material, leading to a form of quantum confinement. In this manner, quantum dots can be thought of as a form of artificial atom whose quantized band structure, and thus properties, depend strongly on the size of the nanocrystal as well as on its material composition. Quantum dots are most simply manufactured by colloidal chemistry techniques, giving good control over both their chemistries and size distribution.

The most obvious application of quantum dots is as efficient fluorescent emitters (and hence good absorbers of light) for devices such as photovoltaic cells and LEDs as well as for biotagging purposes. In all these cases, they possess several advantages over traditional organic dyes in that their quantum yield can be high, they are very photostable, their emission wavelength depends very strongly on the diameter of the dot (leading to tunable emission), and they exhibit a relatively sharp bandwidth. By manufacturing quantum dots with a core–shell structure (the classic example being CdSe–CdS) it is possible to further increase photostability and excitonic lifetime since the exciton is shielded from the local environment. In addition to optical applications, quantum dots have also been used in single-electron transistors and have been mooted as possible qubits for quantum computing applications.

However the sensor operates, whether by direct, indirect, labelled or label-free means, we may also classify optical sensors by their structures, which are typically wave guides, resonators, or a combination of both. Light is confined within these structures as bound electromagnetic modes with strongly enhanced optical fields compared to free photons. These enhanced fields maximize the light–matter interaction giving sensitivity to the presence of the analyte. In practice, the optical fields of such structures are highly localized and decay exponentially externally, and it is within this evanescent field region that the analyte is sensed. Mode propagation is perturbed in the case of complex permittivity sensing, whereas in the case of direct luminescence or labeled detection, the strong evanescent fields may excite luminescence from the analyte/tag which can be detected externally, or be coupled back into the guide and detected on-chip. Examples of waveguide geometries include optical and photonic crystal fibres, as well as planar and integrated optics structures. An important extension to waveguide sensing relates to interferometer structures, where light propagates through both sensing and reference branches of a guide before recombining and interfering to give the output signal. The presence of the reference branch reduces or even eliminates common mode noise and effects due to nonspecific adsorption.

Resonator-based sensors utilize a similar transduction principle to waveguide sensors since their bound optical mode's evanescent field is similarly sensitive to the presence of an analyte, but they have a greater effective light–analyte interaction length as bound photons recirculate within the resonator and thus the sensing region. This multiple pass operation increases sensitivity and a corollary is that small, even nanoscale, resonators compare in sensitivity with waveguide sensors having much larger sensing area. It is then possible to imagine fabricating large arrays of individual resonator sensors in a compact area on chip allowing massively parallel (bio)chemical detection. Resonator geometries include cavities such as Fabry–Pérot structures and photonic crystal defects, as well as structures supporting whispering-gallery modes such as microrings, microdisks, microtoroids, and microspheres. Metal nanoparticles supporting localized surface plasmons may similarly be considered as a form of resonator. The key nanotechnological advance is to utilize the optical field enhancement resulting from confined structures to substantially reduce sensing volumes and enable massively parallel compact sensing arrays.

5.2.4.2 Nanophotonics

In common with other areas of science, the advances in fabrication, measurement, and characterization associated with nanotechnology have had a significant effect in the optics and photonics field, creating the burgeoning discipline of nanophotonics. A report from the MONA consortium [77] examined the expected advances and likely impact of nanophotonics, creating a roadmap for the key optical technology areas of displays, photovoltaics, imaging, lighting, data/telecoms, optical interconnects, and sensors. The aspects of nanophotonics recommended for sensor development were, in order of increasing impact, high refractive index contrast semiconductor nanostructures, microstructured fibres, fluorescent quantum dots/wires, and plasmonics.

Advances in high refractive index contrast semiconductor nanostructures will have greatest impact in integrated optics sensing applications, enabling compact devices composed of dense arrays of high-Q resonator sensors. On-chip integration will allow these resonators to be coupled to waveguides and/or interconnects and thence to processing elements enabling the rapid local detection, speciation, and classification of complex mixtures of analytes. For remote sensing applications, optical fibres provide economic, facile, and robust signal delivery to areas of interest, potentially in hostile environments, and as such are a platform for a number of sensor technologies. Microstructured fibres utilizing Bragg gratings have long been used in the telecoms sector, and the use of photonic crystal fibres extends their application to optical sensing. Photonic crystal fibres guide light in a totally different fashion to conventional fibres, relying upon a photonic bandgap to prohibit light propagation perpendicular to the fibre axis. The bandgap arises from a periodic array of holes running along the length of the fibre, and by careful design, these may also be used to channel fluid analytes through the fibre. The resulting physical overlap between the optical fields of the guided mode and the fluid allows sensing throughout the entire length of the fibre, increasing light–analyte interaction and thus sensitivity. This observation highlights the importance of integrating analyte delivery and handling within the sensor platform, suggesting that microfluidics will have become increasingly important in optical sensors especially when applied to lab-on-a-chip systems. For indirect sensing applications, fluorescent quantum dots and wires are increasingly used as labels as they exhibit high quantum efficiency and brightness, increased photostability, and spectrally narrow emission bands when compared to traditional organic fluorescent dyes. Advances in synthesis to allow better control of particle size distribution, surface functionalization, and structure will lead to enhanced tunability, better compatibility, temperature stability, and lifetime.

5.2.5   Plasmonics

Plasmonics has attracted much scientific interest and has great potential impact in a wide range of application areas, not just in sensing. Surface plasmons (SPs) are electromagnetic modes which exist at metal–dielectric interfaces, consist of longitudinal surface plasma oscillations coupled with an electromagnetic surface wave, and can be excited by incident electromagnetic radiation in a process known as surface plasmon resonance (SPR). SPs exist in two forms; propagating surface plasmon polaritons (SPPs), and nonpropagating localized surface plasmons (LSPs) [76, 79].

5.2.5.1   Surface plasmon polaritons (SPPs)

SPPs are essentially bound, and thus nonradiative, waveguide modes whose fields are strongly localized at an interface, decaying exponentially into the bounding media: $\approx \lambda$

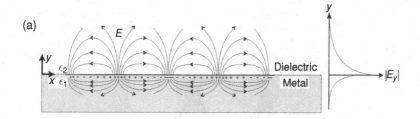

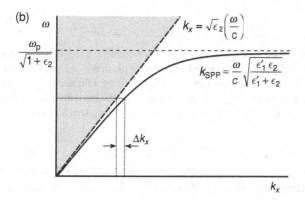

Figure 5.12 (a) The field and charge distributions associated with the SPP mode of a planar metal–dielectric interface. ϵ_1' and ϵ_2 are the real parts of the dielectric constants of the metal and dielectric respectively, and E is the electric field. (b) The SPP dispersion curve for a free-electron metal surface. The gray region represents the range of frequencies and wavevectors accessible to light propagating in the dielectric above the metal, and is bounded by the light-line. Note that k_x is the in-plane wavevector, k_{SPP} the SPP wavevector, ω the circular frequency, c the speed of light, and Δk_x the extra wavevector required to couple the SPP to a free photon.

into the dielectric and ≈ 10–20 nm into the metal for optical wavelengths, depicted in Figure 5.12(a) for a single planar metal–dielectric interface. This interfacial confine-ment enhances SPP field strengths by several orders of magnitude with respect to the incident light field, enabling strong interaction with nearby matter. Since SPP modes are bound to the interface, their in-plane momentum is enhanced (Figure 5.12(b)) and extra momentum must be provided to free photons to excite nonradiative SPPs; this is usually achieved using prism or grating coupling techniques. SPPs are transverse magnetic (TM) modes since there has to be a component of the electric field normal to the interface to generate the necessary surface polarization charge. SPPs have long been studied due to their utility (sensing applications in particular benefit from the enhanced fields [74]) and involvement in fundamental physical processes. Interest has been redoubled by the ability to pattern metal surfaces with nanometer-scale resolution, thereby altering the nature of SPP modes and revealing new insights into their physics, a process aided by advances in nanophotonics, such as the introduction of the scanning near-field optical microscope. These developments allow greater control over the generation and prop-agation of SPPs, with studies revealing their key role in mediating such processes as

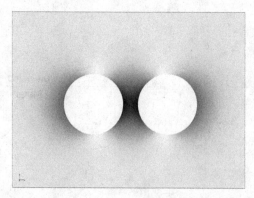

Figure 5.13 The coupled dipole LSPR of a pair of 200-nm-diameter silver nanoparticles separated by 50 nm showing the localized and enhanced fields in the gap region. The plot depicts the normalized electric field distribution calculated using a finite-element model, the black and white regions corresponding to high and low field strengths respectively. The frequency of this resonance is 342.7 THz, equivalent to a wavelength of 875 nm.

surface enhanced raman scattering (SERS) [75] and the enhanced transmission of light through subwavelength hole arrays in metal films. The role that SPP excitation on electrodes plays in reducing organic LED efficiency has also been addressed; interestingly, the solution may be to increase the generation of SPPs, and recover the energy through Bragg scattering.

5.2.5.2   Localized surface plasmons (LSPs)

LSPs are nonpropagating plasmon modes which are supported by, and localized upon individual metal nanostructures such as particles, shells, tips, holes/wells in metal films, or defects in plasmonic crystals. Structure sizes typically range from 10 nm to 100 nm for LSP resonances (LSPRs) in the visible. In contrast with bound SPPs that require a momentum-enhancement coupling mechanism, LSPs can be excited by and can scatter free photons, simplifying their application in devices. The spectral location and strength of the LSP resonance depends on the size and shape of the nanostructure, as well as its material properties and the local dielectric environment. The latter dependence indicates that adsorbates on metal nanostructures may be sensed by observing LSPR changes induced by the local permittivity change. More significantly, the dependence of the LSPR on the geometry of the nanostructures means that resonance wavelengths can be tuned lithographically, giving spectroscopic control. As with SPPs, the optical fields of LSPs are highly localized and enhanced over those of the incident light which has led to spectroscopic applications such as SERS and tip enhanced Raman scattering (TERS). Knowledge of how LSPRs depend on nanostructural geometry has increased understanding of the SERS mechanism itself, which allied with progess in nanofabrication techniques has allowed the rational design of substrates designed to maximize SERS. One key observation is that local electromagnetic field strengths are significantly enhanced in the nanoscale gaps between metal nanoparticles in close proximity, leading to "hot-spots" of Raman signal for nanoparticle dimers, see Figure 5.13.

5.2.5.3 SP-supporting materials

Strictly speaking, SPs are supported at boundaries between materials with positive and negative real components of the dielectric constants. The most common materials exhibiting a negative real dielectric constant are metals, and in the visible range of the electromagnetic spectrum, the noble metals, silver and gold, have the best performance due to their relatively low optical absorption. With increasing wavelength, most metals become more ideal in response as the real part of their dielectric constant becomes increasingly negative and the fields become increasingly excluded from the metal. In general, interface-localized SPs exist only at frequencies close to the plasma frequency of the supporting material. In the terahertz regime, the SPP modes of metals are known as Sommerfeld–Zenneck waves as they become increasingly delocalized and behave as photons propagating in the half-space above the dielectric. These waves do not have the subwavelength confinement of SPPs in the visible regime, and so the advantages associated with plasmonics are lost. Highly localized terahertz SPPs can be obtained by replacing the metal by a heavily doped semiconductor with a plasma frequency in the infrared (rather than in the ultraviolet, which is the most common case for metals). An alternative method which is currently receiving much interest is to pattern the surface of an ideal metal with periodic arrays of subwavelength holes or grooves. Considering the patterned surface using an effective medium theory approach, one can see that these holes allow the field to "penetrate" into the metal providing the evanescent field required to allow strong surface localization, and the resulting modes are known as "spoof plasmons."

5.2.5.4 Plasmonics applications in sensing

The key properties of SPs for sensing applications are their strongly localized optical fields and consequent field enhancement with respect to the incident light, which promotes increased interaction with the local environment. These properties have been utilized for many years but are finding new life in the field of plasmonics, driven by the ability to fabricate metallic nanostructures which localize light at spatial dimensions smaller then the wavelength of light. This study of SPPs propagating upon subwavelength scale structures takes advantage of the lack of a dimensional cutoff for SP propagation (in contrast to conventional optical waveguide modes which cannot propagate on structures with dimensions less than the cutoff, typically $d \approx \lambda/2$). Plasmonics has thus the potential to allow increased component densities for the miniaturization of integrated optics and photonic devices, to empower near-field optics and perhaps improve spatial sensitivity for diagnostics, sensing, and characterization.

The strongly localized and enhanced fields associated with SPs make them particularly sensitive to the optical parameters of the interfacial region, and many studies use the excitation of SPPs as a powerful surface-specific characterization tool. The most common process involves measuring intensity, angle, or wavelength shifts in SPR features observed in sample reflectivity; either using attenuated total reflection, or diffraction grating geometries. This technique has been used extensively to measure the optical constants and surface roughness of metals as well as thin overlayers and adsorbates [79]. Probably the most widespread use of SPR, however, is in sensing applications, where a change in the interface parameters is detected [74]. Examples include the original application of SPR sensing to gases [78], as well as biomaterials and organic condensates. In a similar manner SPR has even been used to monitor electrochemical reactions. In common with other sensing techniques, the challenge lies in achieving chemical specificity, and not in the detection threshold. A solution to this problem is to use a biorecognition approach such that only the analyte binds to the sensing surface.

LSPR sensing typically utilizes metal nanoparticles which are functionalized to interact with the analyte of interest [80]. These can be simply dispersed in solution, or fabricated in arrays using conventional techniques such as e-beam or nanoimprint lithography or by innovations such as nanosphere lithography or microcontact printing. The analyte is detected spectroscopically by observing the shift in LSPR upon analyte binding. In the solution case, this process can be enhanced by causing nanoparticles to agglomerate upon analyte binding, leading to large shifts in LSPR spectra. Arrays of nanostructures are more naturally suited to device applications, and the ability to fabricate and separately functionalize multiple sizes and shapes of nanoparticles means that massively parallel sensing strategies are possible. This is a great advance over standard SPP-based sensing since very small area devices result. Traditional SPR sensing can be enhanced by coupling the analyte with nanoparticles to increase the perturbation of the SPR. If the nanoparticles additionally support LSPs, electromagnetic coupling between the LSPs and the SPP mode of the interface gives greater SPR shifts. Other geometries such as arrays of nanoholes and nanowells have also been used for LSPR sensors, and these have the advantage that the enhanced optical fields are more confined and can offer higher sensitivity.

5.2.6   Electrochemical sensors at the nanoscale

5.2.6.1   Electrochemical sensors

The general mechanism of chemical sensing is based on molecular recognition. Upon recognition, different strategies such as electrochemical, optical, mass, and thermal methods can transduce the signal into a digital readout. Capable of being easily embedded and integrated into electronics, electrochemical sensors provide accurate, fast, and inexpensive analysis, which can be applied in the areas of healthcare, environmental monitoring, food packaging, and in industrial applications [81].

Electrochemical sensors measure the current, voltage, and/or conductance changes that result from chemical reactions which transfer or separate electric charge. Amperometric and voltammetric sensors measure the currents generated by heterogeneous electron transfer reactions such as the oxidation or reduction of electroactive analyte species and, in these measurements, the current is proportional to the analyte concentration. Potentiometric sensors, which include ion-selective field effect transistors (ISFETs, Figure 5.8) and ion selective electrodes (ISEs), relate the potential to the analyte concentration when the electrode reaction is at equilibrium (i.e., with no current flow through the electrode). In general, the potential difference shows a linear relationship with the logarithm of the analyte activity, as given by the Nernst equation. Conductometric sensors are composed of resistive and/or capacitive sensors that measure resistivity changes due to chemical reactions, or capacitance changes resulting from dielectric constant modification.

5.2.6.2 Impact of nanostructures on sensing

A number of research groups have started to explore an integrated approach combining electronics and biology to build biosensors at the nanoscale. Nanoelectrodes promise access to microenvironments, such as cells, which are not accessible to larger electrodes and these analytical capabilities allow the detection of single cell secretion [82] and single molecules [83], measurements of local concentration profiles, detection in microflow systems, and *in vivo* monitoring of neurochemical events by detection of stimulated dopamine release [84]. Devices based on nanowires are emerging as powerful platforms for the direct detection of biological and chemical species, including low concentrations of proteins and viruses [85].

There are several benefits of using nanoscale electrodes in electrochemical sensors. Current density has been shown to increase from the center to the edge of a microelectrode while resistance decreases. At short scan times this means that the ohmic drop is proportional to the radius of the electrode while at long scan times it reaches a constant value, so at short scan speeds nanoelectrodes show a much smaller ohmic potential drop than macroelectrodes [94]. This reduced ohmic drop distortion allows the detection of electrochemical reactions in poorly conducting media, even in the absence of a supporting electrolyte. Similarly double layer capacitances are proportional to electrode area, and electrochemical cells with nanoelectrodes exhibit small RC time constants, allowing high scan rate voltammetric experiments that can probe the kinetics of very fast electron transfer and coupling reactions or the dynamics of processes such as exocytosis [86].

Figure 5.14 illustrates the different cyclic voltammograms generated using either macroscopic or nanoscopic electrodes, which are determined partially by the mass transport of electroactive compounds to the electrodes. Planar diffusion is the main mass transport mechanism when using traditional macroscopic electrodes and redox reactions are limited by mass transport. With macroscopic electrodes the cyclic voltammogram is affected by the ohmic drop despite the presence of the supporting electrolyte. Both effects contribute to the peak shape of the cyclic voltammagram shown in Figure 5.14(a). As the diameter of the electrode is decreased to the nanomenter range radial diffusion becomes dominant over planar diffusion, as shown in Figure 5.14(b).

In addition to mass transfer rates, the voltammetric response of an array of nanoelectrodes depends on the scan rate of the experiment and this gives rise to three limiting cases [94]. At very high scan rates the diffusion layers are thin and extend linearly from the nanoscale electrodes. At lower scan rates each electrode develops a radial diffusion field. If the distance between the individual electrodes is sufficiently large to prevent the radial diffusion fields from merging, then the voltammogram becomes sigmoidal [94]. At very low scan rates the diffusion layers at each nanoelectrode merge to yield a net linear diffusion field and the behavior of the nanoelectrode is comparable to that observed at very high scan rates. Theoretical modeling shows that radial diffusion provides more rapid mass transfer, enabling faster electrochemical measurements that support the analysis of reactions with rapid kinetics [87]. Selection of the scan rate and an array design

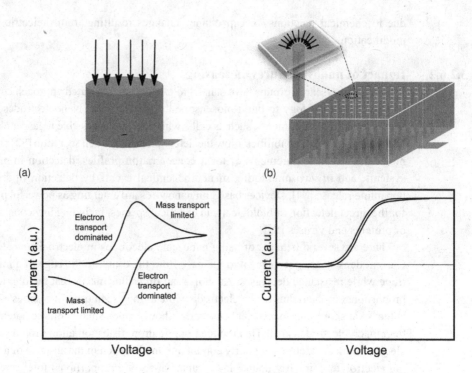

Figure 5.14 (a) Schematic of a macroscopic electrode indicating mass transfer via planar diffusion and the corresponding cyclic voltammogram for a redox reaction. (b) Schematic of a nanoelectrode array with enlargement of a single nanoelectrode to illustrate local radial diffusion, together with its steady state voltammogram at a medium scan rate.

that gives sufficient distance between nanoelectrodes are key to optimal performance by nanoelectrode arrays.

An additional benefit of using nanoelectrode arrays run at scan rates where radial diffusion dominates is that they show excellent signal-to-background ratios. The background signal is predominantly caused by a double-layer charging current at the electrode–solution interface and is proportional to the area of the conductive portion of the electrode. For a macroscale electrode, the conductive area is equal to the total geometric area, but for nanoelectrode arrays, the conductive area is a small fraction of the total geometric area. When comparing a planar electrode to a nanoelectrode array with identical geometric area, the signal amplitude is the same but the much lower conductive area of the nanoelectrode array enables a decrease in background of several orders of magnitude. This strong signal-to-background ratio ensures that nanoarrays show much greater sensitivity [95]. Detection limits for nanoscale biosensors typically lie in the femtomolar concentration range and, as this is likely limited by analyte transport rather than signal transduction [21], it remains possible that ultimate sensitivities have not yet been reported for nanoelectrode arrays.

Electrode functionalization is of key importance for increasing the range of analytes amenable to electrochemical analysis [89]. However, appropriate functionalization schemes must develop low ion concentrations during the electrochemical reaction since ionic screening can diminish the response and significantly increase the incubation times for nanoelectrodes [90]. Nanoelectrode arrays should allow the implementation of massively parallel multianalyte measurements, or statistically significant repetitions of a single analytical measurement [88], but the realization of very-high-density sensing arrays of individually addressable nanoscale elements assumes that functionalization chemistries can be developed that are suitable for performing molecular recognition of thousands of different species in parallel. Electrochemical sensors continue to develop as the technology is amenable to miniaturization and can provide accuracy and sensitivity with simple self-contained instrumentation [91]. The tremendous potential of nanoscale biosensors for highly sensitive electronic detection of biomolecules has already been demonstrated in genomics [92] and proteomic analysis [89].

5.3 Sensor signal processing

5.3.1 Nanoscale enablers for signal processing

Figure 5.3 illustrates the complexity of a state-of-the-art smart sensor that is capable of sensing, signal processing, controlling actions, and communicating with other smart sensors and information systems. This kind of autonomous sensor enables the creation of efficient sensor networks with embedded intelligence to process information and optimize the transfer of data between sensor nodes. A modular approach based on the capabilities of MEMS and CMOS technologies enables the deployment of smart sensing in various real world systems.

In Chapter 4 we highlighted the impact of nanoelectronics upon future computing paradigms and discussed possible nanoelectronic architectures. In Section 5.2, we have shown that nanoelectronics will create new transducer principles and allows us to arrange transducers and related signal processing elements into new kinds of energy-efficient and distributed configurations.

Another important conclusion of Chapter 4 is that while future nanoscale electronics may provide us with superior solutions for some particular computational tasks, they are not always an optimal replacement for CMOS-based digital electronics. Sensor signal processing provides several examples of computational tasks that are good candidates for solution by nanoscale arrays of computing elements, such as filtering, Fourier transformation, signal coding, analog-to-digital conversion, and pattern recognition.

Here we focus on some of these nanoscale opportunities in more detail. Firstly, we will discuss signal amplification, filtering, and primary coding using nanoscale components. Secondly, we will describe possible methods for working with sensors that have very low noise margins and even ways to exploit noise in signal processing. The use of large arrays of similar components – made using bottom-up fabrication processes – for sensing and signal processing opens new paradigms that will be discussed. We present ways in which modern theories of neural computing, artificial intelligence, and machine learning can be applied to the development of nanoscale sensing and signal processing. Finally, as a special technology, we discuss optical signal processing based on plasmonics.

5.3.2   Power amplification and signal coding

Transducers are either active and self-powering or passive parametrically modulating devices. Active transducers typically convert input signal energy into electrical energy, e.g., piezoelectric materials convert mechanical energy into electrical energy. Passive devices are characterized by some parametric change (e.g., capacitive, resistive, or inductive change) in the electrical measurement system and external electrical power is needed to generate a measurable electrical signal. In both cases the signals of these nanoscale devices need to be amplified by a low-noise preamplifier that is matched to the impedance of the transducer, i.e., the real part of the transducer impedance needs to be equal to the input impedance of the preamplifier [96]. The weaker the primary signal, the nearer the preamplifier needs to be placed to the transducer. Integration of nanoscale transducers with measurement electronics is crucial for good measurement (Section 5.2.3). Nanoscale transducers and preamplifiers typically have large $1/f$ noise at low frequencies as the $1/f$ noise is known to scale inversely proportionally with some effective surface area of the device, such as a junction or material interfacial area. The problem of low-frequency noise can only be overcome by some modulation scheme operating above the $1/f$ noise corner frequency (see Figure 5.4).

In general there are two different strategies for integrating nanoscale transducers into measurement electronics. We can use a CMOS integrated circuit wafer as a substrate and integrate the nanoscale transducers on specific locations near the preamplifier transistors [116, 121], or we can use the same nanoscale technology to create both the transducers and the power amplifying devices, e.g., nanowire transducers and transistors.

Besides analog signals, different pulse modulation schemes, such as frequency, pulse width and pulse density modulation have been used for sensor signal coding. Pulse modulated signals can be generated by different parametrically controlled oscillators and easily converted into digital form (see [4]). The advantage of frequency and pulse modulated signals is that they are robust against noise, attenuation, and disturbance, but they require synchronization based on a reference clock signal.

The feasibility of using spike modulated signals has been studied [97, 98, 99]. Spike modulation is based on computational elements known as spiking neurons that can be modeled as a leaky integrate-and-fire circuit governed by the equation $\tau \dot{u} = -u + R i_{in}$ where u is the state variable (voltage), $\tau = RC$ is the time constant of the leaky integrator, R is the discharge resistance, and i_{in} is the input current. Figure 5.15 shows a generic architecture for a spiking neuron that consists of a summing unit that buffers the input signal from transducers or other connected spiking neurons, a leaky integrator, and a threshold element that generates the output spike and resets the integrator when the threshold is passed. Furthermore, the spiking neuron can be trained by adjusting synaptic weights, the integration coefficient, and/or threshold levels.

Signal coding can be based on the spike rates: the spike count and density at the level of a single neuron or a neuron population. However, spiking neuron dynamics allows much more complex coding schemes (for a comprehensive review see [97]). The timing of individual spikes, phase coherence of incoming signals, correlation and synchrony of different inputs, and possible reverse correlation are mechanisms that can be used

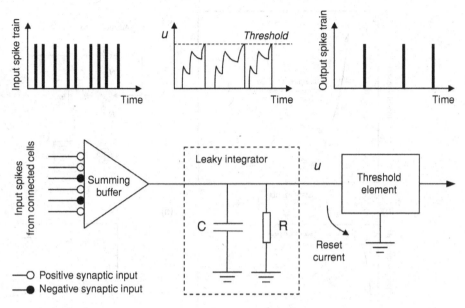

Figure 5.15 Simple model of a leaky integrate-and-fire circuit that can be used to model a spiking neuron.

for signal processing and coding using spike trains. The algorithmic possibilities in the population levels of spiking neurons are various and provide very fast reactions to external stimuli. Spiking-neuron-based components are promising for transducers and signal processing, and nanotechnologies provide new means to implement synaptic weighted integrators and thresholds.

Spiking neurons have been implemented using CMOS technologies, however, these implementations have been complex and require large areas of a silicon chip. In addition, CMOS-based neural networks are constrained by the number of connections that is feasible between neurons in comparison with any mammalian nervous tissue (see also the discussion in Chapter 4). A nanoscale implementation of a spiking neuron based on a MOSFET with multinanodot floating-gate arrays has been proposed [98]. This spiking neuron circuit integrates the essential functionality into one single component, and these building blocks can then be used to create more complex circuitry. There may be other interesting implementations of spiking neuron building blocks based on nanodots, nanowires, nanomechanical resonator arrays, or memristor elements [100, 101, 102, 103, 104, 105]. In summary, nanostructures provide a feasible way to implement an artificial neuron, and, conversely, spiking neuron circuits can create suitable architectures for nanoeletronics.

5.3.3 Noise enhanced signal processing

Noise in various nonlinear threshold systems, such as neurons, Josephson junctions, and Schmitt triggers, can improve signal detection [106, 107]. The phenomenon is

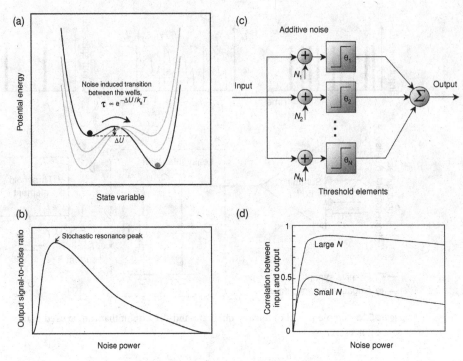

Figure 5.16 (a) Potential energy graph of an externally driven Duffing oscillator, $\ddot{x} + \gamma \dot{x} + kx \pm k_3 x^3 = A \cos \omega t$. The external drive signal $A \cos \omega t$ is below the switching threshold and does not induce switching between the stable state without additional noise energy. (b) SNR of the output signal in the subthreshold system of the Duffing oscillator as a function of noise power. (c) Generic architecture for suprathreshold stochastic resonance in the system of N parallel noisy threshold devices, according to [107, 108]. N_1 and N_2 are noise additive terms and θ_1 and θ_2 are the thresholds. (d) Correlation of input and output signals as a function of noise power for a small and a large number of the parallel threshold devices shown in (c).

known as stochastic resonance and is defined as the condition where noise enhances the input-to-output performance of a nonlinear system.

Figure 5.16(a) illustrates an example of a nonlinear system, an externally driven Duffing oscillator, in which noise can lower the threshold in detecting a signal. When driven at larger oscillation amplitudes, the nonlinear dynamics of the Duffing oscillator can form bistable states, i.e., the two potential wells with stable states indicated by the circles. As the external signal is detected by the change of the state of the oscillator, additional noise can induce a transition of the system state x for lower amplitudes of the external signal. Thus the noise enhances the sensitivity of the system to the external drive signal $A \cos \omega t$. The figure-of-merit of stochastic resonance is the improvement in the correlation between the input and output signals in the system, sometimes measured as the SNR at the output. It can be shown both theoretically and experimentally that there exists an optimal noise power to achieve the best SNR performance and the highest input–output correlation, see Figure 5.16(b). The key challenge in the use of subthreshold stochastic resonance is that it requires intelligence to tune the system in terms of the point of operation of the nonlinear system and the noise power.

Noise enhancement is not only related to subthreshold systems. Furthermore, precise tuning is not critical in ensembles of threshold elements when their response is averaged [108]. Figure 5.16(c) presents an architecture of a stochastic pool network that consists of N nearly identical threshold elements with additional noise connected in parallel with respect to the input signal. The output of the threshold elements is summed or averaged to get the final output signal. It has been shown that such systems work also for aperiodic input signals and that the signal does not need to be entirely subthreshold. The effect has been called suprathreshold stochastic resonance (SSR) to distinguish it from single-device subthreshold systems. The effect is maximized when all the threshold values are set to the signal mean [107]. Figure 5.16(d) shows the input-to-output correlation for an SSR system as a function of noise power for a small and a large number (N) of threshold elements. Incorporation of additive noise in a multicomponent detection system improves the system's capability to detect weak signals.

Stochastic resonance has been studied in the context of low SNR systems and distributed sensor networks. It is also very well understood that stochastic resonance is a general principle for enhancing sensory systems and neural signal coding in nature. Thus the relevance to the sensors and signal processing implemented using nanoscale building blocks is obvious. Both subthreshold stochastic resonance and SSR can be used to implement nanodevices [107, 109, 110, 111, 112, 113].

The observation of stochastic resonance in bistable nanomechanical resonators has been reported by Badzey and Mohanty [109]. Double-clamped, single-crystal silicon beams were driven into transverse oscillation with a roughly 25 MHz drive signal. Without an external noise or modulation signal, the system could be described by the Duffing equation with bistable distinct states as shown in Figure 5.16(a). Adding low-frequency modulation (0.05 Hz square wave) and thermal noise, they were able to show clear stochastic resonance for optimized noise powers in their low-temperature measurements at 0.3–3 K. The demonstration of subthreshold stochastic resonance may be useful for nanomechanical memory and transducer elements as well as for quantum computing based on resonators. Lee *et al.* [111] have shown subthreshold stochastic resonance using CNT FETs to detect noisy subthreshold electrical signals. Characteristics of these subthreshold CNT FET detectors were applied in simulations to create a pixel-level threshold in an image processing architecture to show stochastic resonance improvement in image rendering.

SSR has been studied by Oya *et al.* [112] and by Kasai and Asai [113]. They applied the architecture of Figure 5.16(c) to two different systems: networks of single-electron box neurons and Schottky wrap gate (WPG) GaAs nanowire FETs. SSR architectures enabled the operation of an array of 50 single electron boxes at room temperature, even though the resonant temperature of the system was 20 K. The Schottky WPG FET system was demonstrated to have a clear noise enhancement with a specific additional voltage noise at room temperature, the noise enhancing the transition of electrons over the Schottky barrier. These two systems show the feasibility of realizing the architectures of Figure 5.16(c) in nanoscale arrays of similar components.

5.3.4   Large arrays of nanoscale components

One of the key opportunities for nanoscale systems is to build huge regular arrays of similar nanocomponents. In principle, these arrays can be used either by accessing the information of the state of single nanocomponents or by accessing the information coded in some collective mode of the nanoscale system. For example, crossbar architectures [114–120, 121] allow the measurement of the state of any junction or crossing point that could be a transducer or memory element. Networks of coupled nanomechanical resonators can be used for creating systems with collective states that can be used to sense and process external signals [122–129].

Crossbar architectures provide solutions for novel memories [114, 121] as well as computing solutions; both field programmable gate array (FPGA) [117] and neuromorphic [116] architectures have been studied. Combining these architectures with transducers is an opportunity to bring signal processing, e.g. adaptive filtering, pattern recognition, and signal classification, to the sensor structure itself. Such an implementation could perform complex signal processing tasks with low energy consumption. Artificial retinae, noses, tongues, and cochleae are natural applications for complex transducer, memory, and signal processing circuitry. Chemical and biochemical sensors can be created with nanoscale crossbars. The integration of nanoscale transducers and computing elements with CMOS circuitry is possible in this configuration [116].

Synchronization is a universal complex dynamical process that is characteristic of weakly coupled nonlinear systems. Studies of regular arrays of electrically or mechanically coupled MEMS/NEMS resonators have shown that the resonators tend to converge into oscillatory phase-locked patterns (i.e., equal frequencies and constant phases). Synchronization in these nonlinear nanoelectromechanical systems creates powerful tools for creating signal processing circuitry. Pattern recognition using a neural network architecture based on self-sustaining MEMS oscillators has been demonstrated by Hoppensteadt and Izhikevich [123]. Their MEMS oscillators are driven in a complex phase-locked loop configuration with a cross-coupling between the oscillators via weighted crossbar connections. Another possibility is to build an ensemble of coupled nanomechanical resonators that are also coupled to external signals. The resonator ensemble is known to synchronize into a collective state that can have interesting internally localized patterns depending on the individual resonators and/or couplings [125–130]. The parametrically driven resonator arrays [127, 131, 132] in particular can be a novel tool for creating sensor signal processing. The parametric changes due to external stimulus can be detected as changing localized patterns in the array of resonators.

Large nanoscale arrays create an opportunity for redundant and fault-tolerant solutions for sensors and signal processing. On the other hand, the engineering challenge of creating complex arrays of nanoscale components that can be exposed to external signals and chemical environments is huge. Using large regular arrays of bottom-up grown nanoscale components gives a unique new opportunity that allows us to derive benefits similar to those from the robust sensing principles used in natural systems (see Section 5.2.2). The mammalian nervous system is organized in a hierarchical way and has

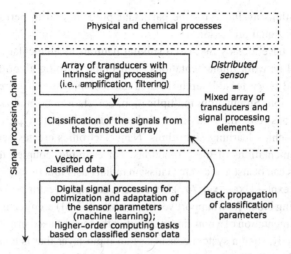

Figure 5.17 Generic sensing architecture based on nanoscale sensing and signal processing in combination with CMOS digital electronics.

a lot of plasticity to adapt to external stimuli (see [133]). Building equally efficient recognition of context and patterns requires novel architectures that can process sensory information in parallel and learn and adapt according to external signals. Large arrays of nanoscale transducers, memory, and computing elements could open a feasible path in this direction.

5.3.5   Machine learning in sensor signal processing

Machine learning methods are a flexible library of different parametric mathematical models [134–137]. These models are generic and can represent many different phenomena depending on model parameters. The key capability of any machine learning method is to learn from example. Based on measured data the model parameters are adjusted so that the model predictions fit the measured data. The parameters discovered can be used to apply the model to new examples. Machine learning methods have advantages for nanoscale signal processing: they can typically be implemented using parallel computing and they are tolerant of any hardware failures.

Figure 5.17 presents a possible architecture of a sensor with signal and data processing capabilities. The system consists of a mixed array of nanoscale transducers and signal processing elements. In this example, the transducer signals are directed to a computing layer (a hidden layer) that is parameterized to classify the signals into classes, such as chemical compounds, or to recognize sequences, e.g., motion or audiovisual patterns. Machine learning relies on the efficiency of computational optimization methods to calculate the parameters of the classifier system. There are ways to implement learning into neural networks but it seems to be more efficient to use the most powerful optimization algorithms available for the particular problem and to perform the computation in a digital signal processor.

We have already discussed spiking neurons as elementary components for building neural network implementations of machine learning algorithms. There are, however, several other feasible architectures and components that are needed for their implementation. Summation, multiplication, correlation, normalization, and different kernel functions (e.g., the Gaussian radial basis function) are typical mathematical primitives for machine learning algorithms. Nanoelectronics creates new possibilities for analog implementations of these functions. For example, coupled nanomechanical resonator arrays can be used to create Gaussian radial basis functions (RBFs) [137]. The Gaussian RBF, $\exp(-\frac{1}{2}|\mathbf{x}_i - \mathbf{x}_j|^2/\sigma^2)$, where $\mathbf{x}$ is an input vector and σ a width parameter, is a building block for a support vector machine (SVM) architecture [136] that is an efficient implementation of a non-linear classifier that can be used in a system such as Figure 5.17. Typically, such a system consists of an input layer that is composed of primary sensor signals, a hidden layer that creates a higher-dimensional hypothesis space, an output layer delivering the result, and a learning bias to train the weights of the system. The generic mathematical form of the output layer is

$$y(\mathbf{x}) = \sum_{i=1}^{n} w_i \, \phi(\|\mathbf{x} - \mathbf{c}_i\|), \tag{5.3}$$

where n is the number of hidden units, w_i are the weights of the summing operation, and ϕ is some nonlinear function of the Euclidean distance between the input vector $\mathbf{x}$ and the set function center $\mathbf{c}_i$. Learning in the system is based on the optimization of parameters w_i and $\mathbf{c}_i$.

Nanowires and nanomechanical resonators can be used to create both the transducers and the hidden layer components [137]. The most advantageous solution is to have similar components in both layers. This physical realization of the transducers and the n-dimensional hidden layer in (5.3) defines the nonlinear functions $\phi(\|\mathbf{x} - \mathbf{c}_i\|)$. For example, nanomechanical resonator arrays can generate a set of Gaussian radial basis kernel functions. Machine learning methods have created a flexible library of mathematical models. We believe that nanoelectronics can be used to build a library of flexible analog computing solutions to implement some of the machine learning algorithms with low power consumption and very fast response in future sensors.

5.3.6   Plasmonics in signal processing

Optical signal processing is a very attractive prospect due to the huge bandwidths that are possible at such high frequencies, however optical waveguides suffer the drawback that diffraction limits the minimum lateral component size to roughly half a wavelength, $\lambda/2$. Typical guide cross-sections that are in the micrometer range thus compare poorly with the $\approx$100 nm possible for electrical interconnects. One way to implement optical frequency signal processing and handling that avoids this limitation is to use plasmonics.

SP-based signal processing and telecom applications possess a significant miniaturization advantage due to the lack of a diffraction limit for appropriate guides; their lateral fields are almost completely guide-confined, without lower cutoff. This contrasts with dielectric guides, which combine large lateral dimensions with considerable field penetration into the cladding media. The usual metal absorption losses (limiting propagation to $\approx 50\,\mu$m for dielectrics, cf. fiber-optics) can be overcome by defining optical elements on a scale much smaller than the propagation length L. Miniature plasmon waveguides are thus ideally suited for incorporation in extremely densely packed integrated optics structures where light can be manipulated and controlled on the subwavelength scale. Plasmonics will not compete with traditional dielectric waveguide technology in telecoms, but will have positive impact in signal processing where the advantage of nanoscale integration can be realized.

The prospect of SPP waveguides enabling high on-chip device densities has already generated spin-off companies, and progress has been made in developing these plasmonic ideas; simple optical elements have been demonstrated that control SPPs freely propagating on extended metal films, e.g., plasmon interferometers. Studies of laterally confined guiding have reported SPP propagation on metal-stripe and wire waveguides, as well as on nanoparticle arrays. For metal-stripe waveguides, attenuation increases as the guide width is decreased, largely due to increased edge scattering, but propagation lengths remain of the order of 10 μm, sufficient to couple plasmonic device elements. Large micrometer-wide metal-stripe guides have been studied extensively, but there are fewer detailed near-field studies of SPP propagation on smaller structures such as nanowires, though far-field observations have been made.

One way to extend the propagation length of SPPs is to utilize symmetrically clad waveguides supporting coupled SPPs. The reduced losses of coupled SPPs arise from the displacement of the electric field from the lossy metal into the bounding dielectric media. As with extended films, theory predicts reduced losses $L \approx 1$ mm for long-range coupled SPs confined to stripe guides and this has been confirmed by experiments at telecoms wavelengths around $\lambda = 1550$ nm. Other relevant plasmonic components have been demonstrated including Bragg gratings, add-drop filters, and switches/Mach–Zender interferometric modulators. An alternative SPP guiding geometry is one that uses the concepts of photonic bandgaps. In this approach, a metal film is patterned with an array of scatterers which exhibits a stop-band for SPPs. Into this array are written defect lines and/or cavities which support SPP propagation, allowing the subwavelength control of these modes. Theoretical studies predict that much simpler geometries such as subwavelength channels cut into metals, and metal gap structures could act as efficient coupled SPP guides, possessing attractive properties such as lossless transmission around 90° bends, long-range coupled mode propagation, and increased mode confinement. Plasmonic devices using these types of guides have been demonstrated. SPP propagation on particle arrays has also been studied, but this approach suffers from short propagation distances, of the order of micrometers. There are intriguing prospects for this approach if the attenuation problem can be overcome, since theory suggests that coherent control may be used to direct the propagation of SPPs.

5.4   Actuation

5.4.1   Principles of actuators in nature and artificial systems

As discussed previously, an intelligent system consists of perception, cognition, and action. Perception and cognition are represented by sensors and signal processing in artificial intelligent systems. Action is delivered by various kinds of actuators that provide an impact on the external environment, typically in the forms of force or motion. Sensing, signal processing, and actuation form a closed loop operation where the measured information is used to control the actuators influencing a physical or chemical process. This kind of control system is typical in industrial processes, robotics, automotive and aerospace systems, and various scientific instruments and is becoming an essential part of consumer goods and electronics. Actuation on the nano- and micro-scales is becoming feasible through the application of new active materials.

There are many biological examples of large macroscopic movements created by structured systems of motor molecules working at the nanoscale. The best characterized are actin and myosin that drive the contraction of muscle tissue, but there has been considerable analysis of the molecular transport proteins kinesin and dynein [139] and the bacterial flagellar rotary motor [140]. These biological motors have inspired the development of nanoscale actuator systems based on proteins that are not normally considered as motors *in vivo* such as the F_0F_1 ATP synthase, RNA polymerase, or even systems developed from DNA molecules [140]. Individual complexes of these biological motor proteins tend to generate movements in the range of nanometers and this is scaled up into macroscale motion at the tissue level by the combination of multiple copies of the motor proteins assembled into ingenious structures that amplify their activities and support their growth and function.

Research to apply nanotechnology to the fabrication of the synthetic equivalent of tissue structure is in its infancy and represents a key challenge to the development of new actuators. There are now a number of nanoscale actuation systems in the literature [140] but commercialization will require a systematic evaluation of their capabilities. Huber *et al.* [5] have presented an excellent framework for evaluating actuators (see Figure 5.18). The performance of actuators can be characterized in terms of several key properties, such as the maximum value of actuation strain ϵ_{max} and the maximum amplitude of the displacement, the maximum value of actuation stress σ_{max}, the maximum value of the strain–stress product $(\epsilon\sigma)_{max}$, i.e., an estimate of the maximum work per unit volume, the strain resolution ϵ_{min}, the frequency response of the actuation strain and stress to the control signal, and the energy efficiency η of the actuator, i.e., the ratio of mechanical work to the input energy.

5.4.2   Nanoscale actuation based on electroactive polymers

There have been numerous attempts to generate synthetic actuation systems that mimic the characteristics of muscle tissue and the leading candidates to date are based on conducting or electroactive polymers (EAPs) [141] as they can be fabricated to exhibit

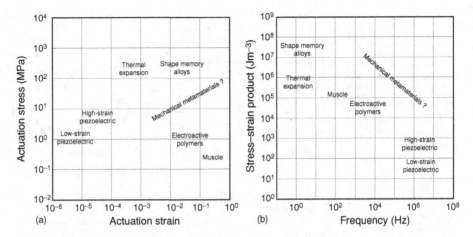

Figure 5.18 (a) Maximum actuation stress and strain for some typical classes of actuator material. (b) Maximum stress–strain product versus frequency for various classes of actuators [5, 138].

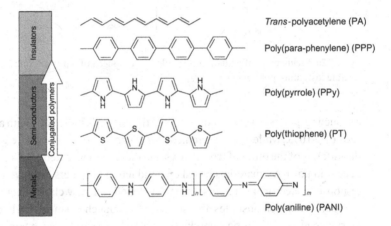

Figure 5.19 Structures of some common conducting polymers.

large strains in response to stimulation. Elastic deformation of polymers or polymer–metal composites can be caused by exposure to a number of stimulators but electrical stimulation is very attractive due to the potential for integration with electrical control systems. Electroactive polymers can be divided into two groups based on their activation mechanism, ionic (involving ion mobility) or electronic (driven by electric fields).

The structures of some common conducting polymers are shown in Figure 5.19. These polymers are typically semiconducting though some can be made as conducting as metals upon doping by chemical or electrochemical methods. The flux of ions and solvents into/out of the polymer matrix during the doping process leads to a volume change that provides the electrochemomechanical action characteristic of conducting polymers. There are many merits of conducting-polymer-based electrochemomechanical actuators when compared to those based on piezoelectrics [144]. The voltage requirement for

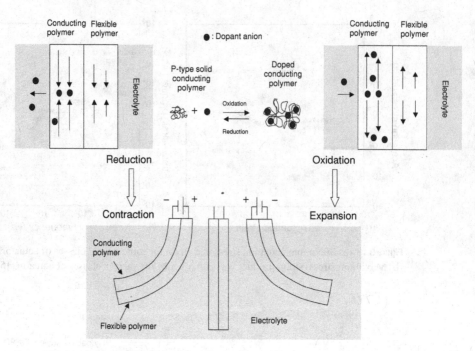

Figure 5.20 Mechanism of actuation by a bilayer composed of a p-type conducting polymer and a flexible insulating polymer.

conducting-polymer-based actuators is of the order of single volts with a fractional extension of >10%, while the requirement for piezoelectric polymers (e.g. poly(vinylidene fluoride)) is of the order of tens of volts and the fractional extension is ca. 0.1%. This new generation of EAP materials based on conducting polymers displays sufficient physical response to electrical excitation to be developed as a new class of actuator.

Figure 5.20 demonstrates the structure and mechanism of EAP bilayer actuators. During oxidation of a p-type conducting polymer, electrolytes containing dopant ions are inserted into the polymer backbone. The polymer expands and stress gradients appear at each point of the interface of the two polymers. During reduction, electrolytes containing the dopant ions are expelled from the conducting polymer backbone. The polymer shrinks and contraction stress gradients appear at each point of the interface of the two polymers making it flex in the opposite direction. An example is a free-standing polypyrrole (PPy) film stuck to one side of a commercially available, double-sided adhesive tape to produce a PPy/flexible-plastic bilayer of dimension 3 cm × 1 cm [145]. A platinum (Pt) wire was connected to the PPy film and functioned as the working electrode. The bilayer was then immersed in an electrochemical cell and various potentials applied using Pt foil as the counterelectrode and a saturated calomel electrode as the reference electrode. It was found that the actuation of the bilayer could be precisely controlled by monitoring and controlling the applied charge rather than the potential or current. This actuation mechanism requires a smooth and uniform conducting polymer film. PPy–gold bilayer actuators that bend out of the plane of the wafer have also been microfabricated

[146, 147]. The microactuators can be used to grab small objects that have geometry suitable for establishing mechanical contact, thus showing the potential to support a number of applications.

Using a Pt foil as the counterelectrode leads to resistive heating and consequent energy loss and so a trilayer planar actuator was designed to improve efficiency. A double-sided adhesive tape (a flexible and insulating polymer substrate) was sandwiched between two layers of conducting polymers and a current was passed between the two conducting layers. The Pt foil counterelectrode was not necessary to manipulate this device and it was reported to be able to pull up to 1000 times its own weight. Trilayer actuators with tactile sensitivity are capable of pushing an obstacle while sensing the required effort [148]. The magnitude and direction of the experimental current control both the rate and direction of the movement. All the actuators discussed above operate in solutions of electrolytes. However, from the point of view of practical applications, solid-state actuators are required. A solid-state actuator based on PPy and a solid polymeric electrolyte has been reported [149]. The solid polymeric electrolyte, poly(epichlorohydrin-co-ethylene oxide), was sandwiched between two PPy layers and provided the ionic contact between the PPy layers.

A number of different nanomaterials have also been developed to support ionic electrochemomechanical actuation. SWCNTs have been reported as an actuator material [150]. CNTs usually exist as tangled bundles that are difficult to dissolve in either aqueous or organic media. However, they can be solubilized and separated in room temperature ionic liquids (RTIL) and CNT-RTIL gels can be prepared [151] that can be laminated together with double-sided adhesive tape to give arrays of nanofiber actuators. SWCNTs have an extremely high Young's modulus and high electric conductivity that supports new characteristics for actuator materials. A solid-state actuator has been fabricated simply through layer-by-layer casting with "bucky gels," a gelatinous RTIL that contains SWCNTs [152]. The nonvolatile RTILs are characterized by their high ionic conductivities and wide range of potential windows, which enable rapid actuation and high electrochemical stability. The performance and durability of the CNT-RTIL gel-based actuators are among the highest reported for low-voltage driven, solid state electromechanical actuators.

Sheets of entangled vanadium oxide (V_2O_5) nanofibers (fibers up to 10 µm long and 10 nm wide) have been shown to contract reversibly on applying an electrical signal when immersed in an aqueous electrolyte [153]. In this case the trilayered actuator is made of two V_2O_5 sheets separated by double-sided adhesive tape. Applying a voltage leads to the intercalation of Li^+ ions in the sheet connected to the anode and their release from the sheet connected to the cathode. One layer swells, while the other one shrinks, leading to the bending of the whole device. The V_2O_5 sheets showed stresses up to 5.9 MPa on application of small potentials (approximately 1 V), which is 10 times higher than that for human skeletal muscles. V_2O_5 is a particularly promising material for use as an actuator as its electronic and ionic properties have already led to its use in microbatteries, electrochromic display devices, and sensors. The composite materials described above show potential as actuators but will need considerable development for full usage, possibly through combining different composites to give structured mechanical metamaterials.

5.4.3   Mechanical metamaterials

Metamaterials have been defined as "Macroscopic composites having a man-made, three-dimensional, periodic cellular architecture designed to produce an optimized combination, not available in nature, of two or more responses to specific excitation" [142]. The three-dimensional architecture is designed to confer a novel function on the metamaterial and this novelty is often used to distinguish a metamaterial from a composite. Most of the scientific literature on metamaterials has focused on the development of novel optical properties with particular interest in the development of cloaking devices or invisibility effects. Kornbluh and colleagues have proposed a category of metamaterials called mechanical metamaterials in which the structure of the material confers an actively controllable mechanical property. The proposed metamaterial includes a deformable structure and a set of activation elements. The connectivity between these elements allows dynamic control of the mechanical properties of this material [143].

Current nanotechnological actuators tend to work in isolation. The challenge is to combine these nanomachines into programmable arrays which can be controlled to give directed mechanical effort on the macroscale. Again one can look to biology for inspiration on how such a metamaterial might be structured. The biological motor proteins actin and myosin represent an extremely well-characterized system for actuation. It has been demonstrated that a myosin protein will bind to an actin filament and, on energy input in the form of chemical hydrolysis, will release and rebind the actin filament giving a movement along the actin filament of approximately 7–12 nm. This energy-dependent nanoscale motion forms the foundation for all muscular actuation [139]. However, despite the dependence on a single type of nanomachine, muscle comes in three types, skeletal, cardiac, and smooth, with each one showing significantly different properties. For example, skeletal muscle generates movement by applying forces to the joints and bones by contraction and is capable of different rates of contraction but is liable to fatigue while cardiac muscle contracts against itself but is highly resistant to fatigue. Smooth muscle is found in the walls of large and small arteries and veins, the bladder, and gastrointestinal tract and differs from the other two muscle types in terms of its structure, function, excitation–contraction coupling, and mechanism of contraction.

Each type of muscle represents a different mechanical metamaterial. The core components are conserved but very different types of actuation are achieved by building these components into specialized three-dimensional structures. These structures show different methods of controlling the actuation, changes in the identities and proportions of the molecules that make up the cellular components of the metamaterials, differences in the strategies used to supply energy to the nanomotors, differences in the structuring of the components to give the mesoscaled actuator (the tissue), and differences in how the system is controlled by external stimuli such as types of nervous excitation. In addition, the muscle-metamaterial develops in situ and is, to an extent, self-repairing.

Muscle tissue is a highly ordered array on nanomachines that provide a diverse set of actuation functions and is an exemplar of the future potential of mechanical metamaterials. However, there are no similar examples in current nanotechnology and there are significant challenges to be met in developing nanotechnological actuation systems that show all the properties of muscle. Actuators based on nanotechnology are in their infancy but we believe that exploitation of some of the biological strategies for their assembly and use will give rise to innovative metamaterials.

5.5   Towards future cognitive solutions

In Section 5.1 we set challenging research questions for nanotechnologies. It is obvious that the development of commercially feasible sensors, signal processing, and actuation based on nanotechnologies is still in a very early phase. Here we have described initial experiments with sensors based on new physical and chemical phenomena. Above all, numerous engineering challenges remain to be solved. In so doing, the research community should bear in mind some key requirements for practical real-life devices, such as feasible voltage levels, low-cost packaging, minimal temperature dependence, optimal complexity, and low overall cost. For example, if the components require very high bias voltages $\sim$100 V and/or ultra-high vacuum packaging, the relevance to sensor development beyond scientific instruments will not be significant.

In Chapter 1 we described applications related to context awareness, mixed and augmented realities, wellness, and healthcare. We discussed solutions and services for global environment and population challenges. Networked sensors and actuators will be enablers for various technical solutions that can build deeper linkages from information networks to our physical environment and that will enrich our capabilities to communicate and to access and share information. There is thus an opportunity for nanotechnologies to build feasible solutions that match with these concepts of future user interfaces, intelligent and responsive spaces, and other artificial cognitive systems. Nanoscale sensing and signal processing have several implications for future systems.

Nanoscale building blocks will not automatically improve the performance of sensors. In many cases miniaturization can even have the opposite effect. An optimal transduction mechanism is matched to the scale of the physical phenomena. We have seen that many new potential sensors derive from chemical and biochemical sensing where the nanoscale transducers create the possibility to derive more detailed information on the observed phenomena. We believe that nanotechnologies offer new possibilities to create nanoscale transducers, memory, and computing elements and to merge these elements together to form an intelligent sensory/actuation system. The same technology, e.g., silicon or ZnO nanowires or CNTs, can be used to create various functional elements for these systems. We have discussed several possible architectures, e.g., coupled resonator arrays, nanowire crossbars, plasmonics, and spiking neuron networks, that can be used for sensing and signal processing.

Macroscopic actuators and mechanical metamaterials that are based on nanoscale physical motion or strain are feasible, as is demonstrated by the fact that life is based on molecular-scale actuators. Nature has been very creative in building actuators in various hierarchical levels of living organisms so that all macroscopic motions of plants and mammalian muscles are based on molecular level actuation. We can either try to mimic these principles or aim to find even more optimal strategies for future technologies.

Microelectromechanical systems have been assembled into arrays of similar functional elements that can be controlled by external forces. These systems can create collective motion patterns that sum to macroscopic dimensions. Nanotechnologies may enable us to embed and assemble these mechanical properties and mechanisms into materials themselves.

The Morph concept is based on the idea of distributed functionality (see Chapter 1). New integration technologies and multifunctional materials enable surfaces and mechanical structures that have embedded functions and intelligence. These are expected to lead to a disruptive change in the low-cost fabrication of large-area functional systems, such as flexible, touch-sensitive, large-area displays with haptic actuation. New manufacturing solutions, such as printed electronics, nanowire networks, and self-assembled functional materials, are steps on the way to this change.

The development of future cognitive solutions will be based on various sensors and actuators together with energy-efficient computing. We believe that these cognitive systems will be implemented on both the global and local levels. The embedding of intelligent sensing and actuation into our everyday environment and artifacts will have the fundamental effect of augmenting the current global mechanism of information exchange, mobile communication, and the Internet, with sensing inputs that in effect not only confer the senses of sight, hearing, smell, taste, and feeling but also sense beyond these biologically defined capabilities. These new linked sensing capabilities will revolutionize our interaction with the digital world but also enrich our connection to the physical world.

References

[1] F. Zhao and L. Guibas, *Wireless Sensor Networks, An Information Processing Approach*, Morgan Kaufmann, 2004.

[2] G.-Z. Yang, *Body Sensor Networks*, Springer, 2006.

[3] T. Ryhänen, Impact of silicon MEMS – 30 years after, in *Handbook of Silicon MEMS Materials and Technologies*, V. Lindroos, M. Tilli, A. Lehto, and T. Motooka, eds., Elsevier, 2009 (in press).

[4] G. Meijer, ed., *Smart Sensor Systems*, Wiley, 2008.

[5] J. E. Huber, N. A. Fleck, and M. F. Ashby, The selection of mechanical actuators based on performance indices, *Proc. R. Soc. Lond. A*, **453**, 2185–2205, 1997.

[6] S. P. Lacour, J. Jones, S. Wagner, T. Li, and Z. Suo, Stretchable interconnects for elastic electronic surfaces, *Proc. IEEE*, **93**, 1459–1467, 2005.

[7] F. N. Hooge, T. G. M. Kleinpenning, and L. K. J. Vandamme, Experimental studies on $1/f$ noise, *Rep. Prog. Phys.*, **44**, 479–532, 1981.

[8] A. Van der Ziel, *Noise in Solid State Devices and Circuits*, Wiley, 1986.

[9] F. N. Hooge, $1/f$ noise sources, *IEEE Trans. Electron Devices*, **41**, 1926–1935, 1994.

[10] J. Chandrashekar, M. A. Hoon, N. J. P. Ryba, and C. S. Zuker, The receptors and cells of mammalian taste, *Nature*, **444**, 288–294, 2006.

[11] E. A. Lumpkin and M. J. Caterina, Mechanisms of sensory transduction in the skin, *Nature*, **445**, 858–865, 2007.

[12] Y. Tu, T. S. Shimizu, and H. C. Berg, Modeling the chemotactic response of *Escherichia coli* to time-varying stimuli, *Proc. Natl Acad. Sci. (USA)*, **105**, 14855–14860, 2008.

[13] S. D. Senturia, *Microsystem Design*, Kluwer, 2002.

[14] J. Monod, J. Wyman, and J. P. Changeux, On the nature of allosteric transitions: a plausible model, *J. Mol. Biol.*, **12**, 88–118, 1965.

[15] V. Sourjik, Receptor clustering and signal processing in *E. coli* chemotaxis, *Trends Microbiol.*, **12**, 569–576, 2004.

[16] M. J. Tindall, S. L. Porter, P. K. Maini, G. Gaglia, and J. P. Armitage, Overview of mathematical approaches used to model bacterial chemotaxis I: the single cell, *Bull. Math. Biol.*, **70**, 1525–1569, 2008.

[17] X.-J. Huang and Y.-K. Chon, Chemical sensors based on nanostructured materials, *Sensors and Actuators B: Chemical*, **122**, 659–671, 2007.

[18] L. G. Carrascosa, M. Moreno, M. Alvarez, and L. M. Lechuga, Nanomechanical biosensors: a new sensing tool, *Trends Anal. Chem.*, **25**, 196–206, 2006.

[19] G. Zheng, F. Patolsky, Y. Cui, W. U. Wang, and C. M. Lieber, Multiplexed electrical detection of cancer markers with nanowire sensor arrays, *Nat. Biotech*, **23**, 1294–1301, 2005.

[20] B. He, T. J. Morrow, and C. D. Keating, Nanowire sensors for multiplexed detection of biomolecules, *Curr. Op. Chem. Biol.*, **12**, 522–528, 2008.

[21] P. E. Sheehan and L. J. Whitman, Detection limits for nanoscale biosensors, *Nano Lett.*, **5**, 803–807, 2005.

[22] W. Yang, P. Thordarson, J. J. Gooding, S. P. Ringer, and F. Braet, Carbon nanotubes for biological and biomedical applications, *Nanotech.*, **18**, 1–12, 2007.

[23] Y. Dan, S. Evoy, and A. T. C. Johnson, Chemical gas sensors based on nanowires http://arxiv.org/ftp/arxiv/papers/0804/0804.4828.pdf.

[24] T. W. Tombler, C. Zhou, L. Alexseyev, *et al.*, Reversible electromechanical characteristics of carbon nanotubes under local-probe manipulation, *Nature*, **405**, 769–772, 2000.

[25] E. D. Minot, Y. Yaish, V. Sazonova, J.-Y. Park, M. Brink, and P. L. McEuen, Tuning carbon nanotube band gaps with strain, *Phys. Rev. Lett.*, **90**, 156401, 2003.

[26] J. Cao, Q. Wang, and H. Dai, Electromechanical properties of metallic, quasimetallic, and semiconducting carbon nanotubes under stretching, *Phys. Rev. Lett.*, **90**, 157601, 2003.

[27] C. Hierold, From micro- to nanosystems: mechanical sensors go nano, *J. Micromech. Microeng.*, **14**, S1–S11, 2004.

[28] C. Stampfer, A. Jungen and C. Hierold, Single Walled Carbon Nanotubes as Active Elements in Nano Bridge Based NEMS, in *Proceedings of the 2005 5th IEEE Conference on Nanotechnology*, IEEE, 2005.

[29] C. Stampfer, T. Helbling, D. Obergfell, *et al.* Fabrication of single-walled carbon-nanotube-based pressure sensors, *Nano Lett.*, **6**, 233–237, 2006.

[30] P. G. Collins, M. S. Fuhrer, and A. Zettl, $1/f$ noise in carbon nanotubes, *Appl. Phys. Lett.*, **76**, 894–896, 2000.

[31] Z. L. Wang and J. Song, Piezoelectric nanogenerators based on zinc oxide nanowire arrays, *Science*, **312**, 242–246, 2006.

[32] X. Wang, J. Zhou, J. Song, J. Liu, N. Xu, and Z. L. Wang, Piezoelectric field effect transistor and nanoforce sensor based on a single ZnO nanowire, *Nano Lett.*, **6**, 2768–2772, 2006.

[33] J. H. He, C. L. Hsin, J. Liu, L. J. Chen, and Z. L. Wang, Piezoeletric gated diode of a single ZnO nanowire, *Adv. Mat.*, **19**, 781–784, 2007.

[34] Z. Wang, J. Hu, A. P. Suryavanshi, K. Yum, and M.-F. Yu, Voltage generation from individual $BaTiO_3$ nanowires under periodic tensile mechanical load, *Nano Lett.*, **7**, 2966–2969, 2007.

[35] X. Wang, J. Song, J. Liu, and Z. L. Wang, Direct-current nanogenerator driven by ultrasonic waves, *Science*, **316**, 102–105, 2007.

[36] W. S. Su, Y. F. Chen, C. L. Hsiao, and L. W. Tu, Generation of electricity in GaN nanorods induced by piezoelectric effect, *Appl. Phys. Lett.*, **90**, 063110, 2007.

[37] J. Zhou, P. Fei, Y. Gao, *et al.*, Mechanical-electrical triggers and sensors using piezoelectric microwires/nanowires, *Nano Lett.*, **8**, 2725–2730, 2008.

[38] F. Patolsky, G. Zheng, and C. M. Lieber, Nanowire-based biosensors, *Anal. Chem.*, **78**, 4260–4269, 2006.

[39] H.-H. Park, S. Jin, Y. J. Park, and H. S. Min, Quantum simulation of noise in silicon nanowire transistors, *J. Appl. Phys.*, **104**, 023708, 2008.

[40] S. Reza, G. Bosman, M. S. Islam, T. I. Kamins, S. Sharma, and R. S. Williams, Noise in silicon nanowires, *IEEE Trans. Nanotech.*, **5**, 523–529, 2006.

[41] J. Kivioja, A. Colli, M. Bailey, and T. Ryhänen, *Double Gated Silicon Nanowire Field Effect Transistors as Charge Detection Based Bio and Chemical Sensors*, to be published.

[42] A. Bid, A. Bora, and A. K. Raychaudhuri, $1/f$ noise in nanowires, *Nanotech.*, **17**, 152–156, 2006.

[43] J. Kong, N. R. Franklin, C. Zhou, *et al.*, Nanotube molecular wires as chemical sensors, *Science*, **287**, 622–625, 2000.

[44] D. Zhang, C. Li, X. Liu, S. Han, T. Tang, and C. Zhou, Doping dependent NH_3 sensing of indium oxide nanowires, *Appl. Phys. Lett.*, **83**, 1845–1847, 2003.

[45] Z. Fan, D. Wang, P.-C. Chang, W.-Y. Tseng, and J. G. Lu, ZnO nanowire field-effect transistor and oxygen sensing property, *Appl. Phys. Lett.*, **85**, 5923–5925, 2004.

[46] Z. Fan and J. G. Lu, Chemical sensing with ZnO nanowire field-effect transistor, *IEEE Trans. Nanotech.*, **5**, 393–396, 2006.

[47] V. V. Sysoev, J. Goschnick, T. Schneider, E. Strelcov, and A. Kolmakov, A gradient microarray electronic nose based on percolating SnO_2 nanowire sensing elements, *Nano Lett.*, **7**, 3182–3188, 2007.

[48] P.-C. Chen, F. N. Ishikawa, H.-K. Chang, K. Ryu, and C. Zhou, A nanoelectronic nose: a hybrid nanowire/carbon nanotube sensor array with integrated micromachined hotplates for sensing gas discrimination, *Nanotech.*, **20**, 125503, 2009.

[49] M. Y. Zavodchikova, T. Kulmala, A. G. Nasibulin, *et al.*, Carbon nanotube thin film transistors based on aerosol methods, *Nanotech.*, **20**, 085201, 2009.

[50] T. Mattila, J. Kiihamäki, T. Lamminmäki, *et al.*, 12 MHz micromechanical bulk acoustic mode oscillator, *Sensor and Actuators*, **A101**, 1–9, 2002.

[51] G. Piazza, R. Abdolvand, and F. Ayazi, Voltage-tunable piezoelectrically-transduced single-crystal silicon resonators on SOI substrate, in *Proceedings of the IEEE Sixteenth Annual International Conference on Micro Electro Mechanical Systems (MEMS-03)*, pp. 149–152, IEEE, 2003.

[52] S. Humad, R. Abdolvand, G. Ho, G. Piazza, and F. Ayazi, High frequency micromechanical piezo-on-silicon block resonators, in *Proceedings of the IEEE Sixteenth Annual International Conference on Micro Electro Mechanical Systems (MEMS-03)*, pp. 39–43, IEEE, 2003.

[53] V. Kaajakari, T. Mattila, A. Oja, and H. Seppä, Nonlinear limits for single-crystal silicon microresonators, *J. Microelectromechanical Systems*, **13**, 715–724, 2004.

[54] A. Cleland, *Foundations of Nanomechanics*, Springer, 2003.

[55] A. N. Cleland and M. L. Roukes, Noise processes in nanomechanical resonators, *J. Appl. Phys*, **92**, 2758–2769, 2002.

[56] K. L. Ekinci, Y. T. Yang, and M. L. Roukes, Ultimate limits to inertial mass sensing based upon nanoelectromechanical systems, *J. Appl. Phys.*, **95**, 2682–2689, 2004.

[57] K. L. Ekinci and M. L. Roukes, Nanoelectromechanical systems, *Rev. Sci. Instrum.*, **76**, 061101, 2005.

[58] A. K. Naik, M. S. Hanay, W. K. Hiebert, X. L. Feng, and M. L. Roukes, Towards single-molecule nanomechanical mass spectrometry, *Nature Nanotech.*, **4**, 445–450, 2009.

[59] W. G. Conley, A. Raman, C. M. Krousgrill, and S. Mohammadil, Nonlinear and nonplanar dynamics of suspended nanotube and nanowire resonators, *Nano Lett.*, **6**, 1590–1595, 2008.

[60] R. He, X. L. Feng, M. L. Roukes, and P. Yang, Self-transducing silicon nanowire electromechanical systems at room temperature, *Nano Lett.*, **8**, 1756–1761, 2008.

[61] X. L. Feng, C. J. White, A. Hajimiri, and M. L. Roukes, A self-sustaining ultrahigh-frequency nanoelectromechanical oscillator, *Nature Nanotech.*, **3**, 342–346, 2008.

[62] A. Colli, A. Fasoli, S. Pisana, *et al.*, Nanowire lithography on silicon, *Nano Lett.*, **8**, 1358–1362, 2008.

[63] B. Lassagne, D. Garcia-Sanchez, A. Aguasca, and A. Bachtold, Ultrasensitive mass sensing with a nanotube electromechanical resonator, *Nano Lett.*, **8**, 3735–3738, 2008.

[64] J. S. Bunch, A. M. van der Zande, S. S. Verbridge, *et al.*, Electromechanical resonators from graphene sheets, *Science*, **315**, 490–493, 2007.

[65] J. T. Robinson, M. Zalalutdinov, J. W. Baldwin, *et al.*, Wafer-scale reduced graphene oxide films for nanomechanical devices, *Nano Lett.*, **8**, 3441–3445, 2008.

[66] K. Jensen, J. Weldon, H. Garcia, and A. Zettl, Nanotube radio, *Nano Lett.*, **7**, 3508–3511, 2007.

[67] K. Jensen, K. Kim, and A. Zettl, An atomic-resolution nanomechanical mass sensor, *Nature Nanotechnology*, **3**, 533–537, 2008.

[68] W. A. de Heer, A. Châtelain, and D. A. Ugarte, Carbon nanotube field-emission electron source, *Science*, **270**, 1179–1180, 1995.

[69] S. Itoh and M. Tanaka, Current status of field-emission displays, *Proc. IEEE*, **90**, 514–520, 2002.

[70] G. Amaratunga, Watching the nanotube, *IEEE Spectrum*, **40**, 28–32, 2003.

[71] A. Ayari, P. Vincent, S. Perisanu, *et al.*, Self-oscillations in field emission nanowire mechanical resonators: a nanometric dc/ac conversion, **7**, 2252–2257, 2007.

[72] C. McDonagh, C. S. Burke and B. D. MacCraith, Optical chemical sensors, *Chem. Rev.*, **108**, 400–422, 2008.

[73] X. Fan, I. M. White, S. I. Shopova, H. Zhu, J. D. Suter, and Y. Sun, Sensitive optical biosensors for unlabelled targets: a review, *Anal. Chim. Acta*, **620**, 8–26, 2008.

[74] J. Homola, Surface plasmon resonance sensors for detection of chemical and biological species, *Chem. Rev.*, **108**, 462–493, 2008.

[75] K. Kneipp, M. Moskovits, and H. Kneipp, eds., *Surface-Enhanced Raman Scattering – Physics and Applications*, Topics in Applied Physics, vol. 103, Springer, 2006.

[76] S. A. Maier, *Plasmonics: Fundamentals and Applications*, Springer, 2007.

[77] Merging Optics and Nanotechnologies Consortium, 2008, *A European Roadmap for Photonics and Nanotechnologies*, Available: http://www.ist-mona.org

[78] C. Nylander, B. Liedberg, and T. Lind, Gas detection by means of surface plasmon resonance, *Sens. Act.*, **3**, 79, 1982.

[79] H. Raether, *Surface Plasmons*, Springer-Verlag, 1988.

[80] M. E Stewart, C. R. Anderson, L. B. Thompson, *et al.*, Nanostructured plasmonics sensors, *Chem. Rev.*, **108**, 94–521, 2008.

[81] E. Bakker and Y. Qin, Electrochemical sensors, *Anal. Chem.*, **78**, 3965–3984, 2006.

[82] C. Amatore, S. Arbault, M. Guille, and F. Lemaitre, Electrochemical monitoring of single cell secretion: vesicular exocytosis and oxidative stress, *Chem. Rev.*, **108**, 2585–2621, 2008.

[83] F. Fan, J. Kwak, and A. J. Bard, Single molecule electrochemistry, *J. Am. Chem. Soc.*, **118**, 9669–9675, 1996.

[84] R. T. Kennedy, L. Huang, M. Atkinson, and P. Dush, Amperometric monitoring of chemical secretions from individual pancreatic beta-cells, *Anal. Chem.*, **65**, 1882–1887, 1993.

[85] Y. Cui, Q. Wei, H. Park, and C. M. Lieber, Nanowire nanosensors for highly sensitive and selective detection of biological and chemical species, *Science*, **293**, 1289–1292, 2001.

[86] C. P. Andrieux, P. Hapiot, and J. M. Saveant, Ultramicroelectrodes for fast electrochemical kinetics, *Electroanalysis*, **2**, 183–193, 1990.

[87] A. J. Bard and L. R. Faulkner, Potential sweep methods in *Electrochemical Methods, Fundamentals and Applications*, second edition, pp. 226–260, Wiley, 2001.

[88] A. J. Bard, J. A. Crayston, G. P. Kittlesen, T. Varco Shea, and M. S. Wrighton, Digital simulation of the measured electrochemical response of reversible redox couples at microelectrode arrays: consequences arising from closely spaced ultramicroelectrodes, *Anal. Chem.*, **58**, 2321–2331, 1986.

[89] E. Stern, J. F. Klemic, D. A. Routenberg, *et al.*, Label-free immunodetection with CMOS-compatible semiconducting nanowires, *Nature*, **445**, 519–522, 2007.

[90] P. R. Nair and M. A. Alam, Screening-limited response of nanobiosensors, *Nano Lett.*, **8**, 1281–1285, 2008.

[91] J. Wang, Electrochemical biosensors: towards point-of-care cancer diagnostics, *Biosens. Bioelectron.*, **21**, 1887–1892, 2006.

[92] J. Hahm and C. M. Lieber, Direct ultrasensitive electrical detection of DNA and DNA sequence variations using nanowire nanosensors, *Nano Lett.*, **4**, 51–54, 2004.

[93] H. Reller, E. Kirowa-Eisner, and E. Gileadi, Ensembles of microelectrodes: digital simulation by the two dimensional expanding grid method. Cyclic voltammetry, *iR* effects and applications, *J. Electroanal. Chem.*, **161**, 247–268, 1984.

[94] I. F. Cheng, L. D. Whiteley and C. R. Martin, Ultramicroelectrode ensembles. Comparison of experimental and theoretical responses and evaluation of electroanalytical detection limits, *Anal. Chem.*, **61**, 762–766, 1989.

[95] V. P. Menon and C. R. Martin, Fabrication and evaluation of nanoelectrode ensembles, *Anal. Chem.*, **67**, 1920–1928, 1995.

[96] Y. Netzer, The design of low-noise amplifier, *Proc. IEEE*, **69**, 728–741, 1981.

[97] W. Gerstner and W. Kistler, *Spiking Neuron Models – Single Neurons, Populations, Plasticity*, Cambridge University Press, 2002.

[98] T. Morie, T. Matsuura, M. Nagata, and A. Iwata, A multinanodot floating-gate MOSFET circuit for spiking neuron models, *IEEE Trans. Nanotech.*, **2**, 158–164, 2003.

[99] A. V. M. Herz, T. Gollisch, C. K. Machens, and D. Jaeger, Modeling single-neuron dynamics and computations: a balance of detail and abstraction, *Science*, **314**, 80–85, 2006.

[100] L. O. Chua and S. M. Kang, Memristive devices and systems, *Proc. IEEE*, **64**, 209–223, 1976.

[101] R. Waser and M. Aono, Nanoionics-based resistive switching memories, *Nature Mat.*, **6**, 833–840, 2007.

[102] D. B. Strukov, G. S. Snider, D. R. Stewart, and R. S. Williams, The missing memristor found, *Nature*, **453**, 80–83, 2008.

[103] J. J. Yang, M. D. Pickett, X. Li, D. A. A. Ohlberg, D. R. Stewart, and R. S. Williams, Memristive switching mechanism for metal/oxide/metal nanodevices, *Nature Nanotech.*, **3**, 429–433, 2008.

[104] M. Rinkiö, A. Johansson, G. S. Paraoanu, and P. Törmä, High-speed memory from carbon nanotube field-effect transistors with high-κ gate dielectric, *Nano Lett.*, **9**, 643–647, 2009.

[105] S. H. Jo, K.-H. Kim, and W. Lu, High-density crossbar arrays based on a Si memristive system, *Nano Lett.*, **9**, 870–874, 2009.

[106] L. Gammaitoni, P. Hänggi, P. Jung, and F. Marchesoni, Stochastic resonance, *Rev. Mod. Phys.*, **70**, 223–287, 1998.

[107] M. D. Donnell, N. G. Stocks, C. E. M. Pearce, and D. Abbott, *Stochastic Resonance – From Suprathreshold Stochastic Resonance to Stochastic Signal Quantization*, Cambridge University Press, 2008.

[108] J. J. Collins, C. C. Chow, and T. T. Imhoff, Stochastic resonance without tuning, *Nature*, **376**, 236–238, 1995.

[109] R. L. Badzey and P. Mohanty, Coherent signal amplification in bistable nanomechanical oscillators by stochastic resonance, *Nature*, **437**, 995–998, 2005.

[110] F. Martorell, M. D. McDonnell, A. Rubio, and D. Abbott, Using noise to break the noise barrier in circuits, in *Proceedings of the SPIE Smart Structures, Devices, and Systems II*, vol. 5649, S.F. Al-Sarafi, ed., pp. 53–66, SPIE, 2005.

[111] I. Lee, X. Liu, C. Zhou, and B. Kosko, Noise-enhanced detection of subthreshold signals with carbon nanotubes, *IEEE Trans. Nanotech.*, **5**, 613–627, 2006.

[112] T. Oya, T. Asai, and Y. Amemiya, Stochastic resonance in an ensemble of single-electron neuromorphic devices and its application to competitive neural networks, *Chaos, Solitons and Fractals*, **32**, 855–861, 2007.

[113] S. Kasai and T. Asai, Stochastic resonance in Schottky wrap gate-controlled GaAs field-effect transistors and their networks, *Appl. Phys. Express*, **1**, 1–3, 2008.

[114] J. R. Heath, P. J. Kuekes, G. S. Snider, and R. S. Williams, A defect-tolerant computer architecture: opportunities for nanotechnology, *Science*, **280**, 1716–1721, 1998.

[115] M. M. Ziegler and M. R. Stan, CMOS/nano co-design for crossbar-based molecular electronic systems, *IEEE Trans. Nanotechnology*, **2**, 217–230, 2003.

[116] Ö. Türel, J. H. Lee, X. Ma, and K. K. Likharev, Neuromorphic architectures for nanoelectronic circuits, *Int. J. Circ. Theor. Appl.*, **32**, 277–302, 2004.

[117] D. B. Strukov and K. K. Likharev, CMOL FPGA: a reconfigurable architecture for hybrid circuits with two-terminal nanodevices, *Nanotechnology*, **16**, 888–900, 2005.

[118] X. Ma, D. B. Strukov, J. H. Lee, and K. K. Likharev, Afterlife for silicon: CMOL circuit architectures, in *Proceedings of the Fifth IEEE Conference on Nanotechnology (2005)*, IEEE, 2005.

[119] A. DeHon, Nanowire-based programmable architectures, *ACM J. Emerging Technol. Computing Sys.*, **1**, 109–162, 2005.

[120] C. A. Moritz, T. Wang, P. Narayanan, *et al.*, Fault-tolerant nanoscale processors on semiconductor nanowire grids, *IEEE Trans. Circuits Syst. I*, **54**, 2422–2437, 2007.

[121] G. S. Snider and R. S. Williams, Nano/CMOS architectures using a field-programmable nanowire interconnect, *Nanotech.*, **18**, 1–11, 2007.

[122] F. C. Hoppensteadt and E. M. Izhikevich, Pattern recognition via synchronization in phase-locked loop neural networks, *IEEE Trans. Neural Networks*, **11**, no. 3, 734–738, 2000.

[123] F. C. Hoppensteadt and E. M. Izhikevich, Synchronization of MEMS resonators and mechanical neurocomputing, *IEEE Trans. Circuits Syst. I*, **48**, no. 2, 133–138, 2001.

[124] A. Pikovsky, M. Rosenblum, and J. Kurths, *Synchronization – A Universal Concept in Nonlinear Sciences*, Cambridge University Press, 2001.

[125] E. Buks and M. L. Roukes, Electrically tunable collective response in a coupled micromechanical array, *J. Microelectromechanical Systems*, **11**, 802–807, 2002.

[126] M. Sato, B. E. Hubbard, A. J. Sievers, B. Ilic, D. A. Czaplewski, and H. G. Craighead, Observation of locked intrinsic localized vibrational modes in a micromechanical oscillator array, *Phys. Rev. Lett.*, **90**, 044102, 2003.

[127] R. Lifshitz and M. C. Cross, Response of parametrically driven nonlinear coupled oscillators with application to micromechanical and nanomechanical resonator arrays, *Phys. Rev.*, **B67**, 134302, 2003.

[128] M. C. Cross, A. Zumdieck, R. Lifshitz, and J. L. Rogers, Synchronization by nonlinear frequency pulling, *Phys. Rev. Lett.*, **93**, 224101, 2004.

[129] M. K. Zalalutdinov, J. W. Baldwin, M. H. Marcus, R. B. Reichenbach, J. M. Parpia, and B. H. Houston, Two-dimensional array of coupled nanomechanical resonators, *Appl. Phys. Lett.*, **88**, 143504, 2006.

[130] N. Nefenov, Applications of coupled nanoscale resonators for spectral sensing, *J. Phys.: Condens. Matter*, **21**, 2009, in press.

[131] E. Goto, The parametron, a digital computing element which utilizes parametric oscillation, *Proc. IRE*, 1304–1316, 1959.

[132] I. Mahboob and H. Yamaguchi, Bit storage and bit flip operations in an electromechanical oscillator, *Nature Nanotech.*, **3**, 275–279, 2008.

[133] L. Lin, R. Osan, and J. Z. Tsien, Organizing principles of real-time memory encoding: neural clique assemblies and universal neural codes, *Trends Neurosciences*, **29**, 48–57, 2006.

[134] J. Hertz, A. Krogh, and R. G. Palmer, *Introduction to the Theory of Neural Computation*, Santa Fe Institute in the Science of Complexity, Westview Press, 1991.

[135] E. Alpaydin, *Introduction to Machine Learning*, MIT Press, 2004.

[136] N. Cristianini and J. Shawe-Taylor, *An Introduction to Support Vector Machines and Other Kernel-Based Learning Methods*, Cambridge University Press, 2000.

[137] M. Uusitalo, J. Peltonen, and T. Ryhänen, *Machine Learning: How It Can Help Nanocomputing*, to be published, 2009.

[138] Y. Bar-Cohen, T. Xue, M. Shahinpoor, J. Simpson, and J. Smith, Flexible, low-mass robotic arm actuated by electroactive polymers and operated equivalently to human arm and hand, in the *Proceedings of Robotics 98: The Third Conference and Exhibition/Demonstration on Robotics for Challenging Environments (1998)*, ASCE, 1998.

[139] D. Voet, J. G. Voet, and C. W. Pratt, *Fundamentals of Biochemistry, Life at the Molecular Level*, second edition, pp. 1072–1114, Wiley, 2006.

[140] A. Ummat, A. Dubey and C. Mavroidis, Bio-nanorobotics: a field inspired by nature, in *Biomimetics, Biologically Inspired Technologies*, Y. Bar-Cohen, ed., Taylor & Francis, 2006.

[141] Y. Bar-Cohen, Artificial muscles using electroactive polymers, in *Biomimetics, Biologically Inspired Technologies*, Y. Bar-Cohen, ed., Taylor & Francis, 2006.

[142] R. M. Walser, Metamaterials: an introduction, in *Introduction to Complex Mediums for Electromagnetics and Optics*, W. S. Weiglhofer and A. Lakhtakia, eds., SPIE Press, 2003.

[143] R. D. Kornbluh, R. E. Pelrine, H. Prahlad and S. E. Stanford, Mechanical meta-materials, International Patent Number WO2005/089176A2, 2005.

[144] R. H. Baughman, and L. W. Shacklette, Application of dopant-induced structure property changes of conducting polymers, in *Science and Applications of Conducting Polymers*, W. R. Salanek, D. T. Clark, E. J. Samuelson, eds., p. 47, Adam Hilger, 1991.

[145] T. F. Otero and E. de Larreta-Azelain, Electrochemical control of the morphology, adherence, appearance and growth of polypyrrole films, *Synth. Met.*, **26**, 79–88, 1988.

[146] E. W. H. Jager, E. Smela, and O. Inganas, Microfabricating conjugated polymer actuators, *Science*, **290**, 1540–1545, 2000.

[147] E. Smela, O. Inganas, and W. Lu, Controlled folding of micrometer-size structures, *Science*, **268**, 1735–1738, 1995.

[148] T. F. Otero and M. T. Cortes, Artificial muscles with tactile sensitivity, *Adv. Mater.*, **15**, 279–282, 2003.

[149] J. M. Sansinena, V. Olazabal, T. F. Otero, C. N. Polo da Fonseca, and M. A. De Paoli, A solid state artificial muscle based on polypyrrole and a solid polymeric electrolyte working in air, *Chem. Commun.*, **22**, 2217–2218, 1997.

[150] R. H. Baughman, C. X. Cui, A. A. Zakhidov, *et al.*, Carbon nanotube actuators, *Science*, **284**, 1340–1344, 1999.

[151] T. Fukushima, A. Kosaka, Y. Ishimura, *et al.*, Molecular ordering of organic molten salts triggered by single-walled carbon nanotubes, *Science*, **300**, 2072–2074, 2003.

[152] T. Fukushima, K. Asaka, A. Kosaka, and T. Aida, Fully plastic actuator through layer-by-layer casting with ionic-liquid-based bucky gel, *Angew. Chem. Int. Ed.*, **44**, 2410–2413, 2005.

[153] G. Gu, M. Schmid, P. W. Chiu, *et al.*, V_2O_5 nanofibre sheet actuators, *Nat. Mater.*, **2**, 316–319, 2003.

6 Future of Radio and Communication

A. Pärssinen, R. Kaunisto, and A. Kärkkäinen

6.1 Introduction

Rapid progress in radio systems since the early 1990s has fundamentally changed all human communication. Mobile phones and the Internet have enabled globalization of both private and business communication and access to information. This has required increased data rates and capacity as well as the means to achieve a reliability similar to that of wireline systems. On the other hand, the importance of local information whether it is the opening hours of a local shop or finding friends in the vicinity will increase the role of short-range radios.

New forms of mobile communication, e.g. location-based services, will increase the need for enhanced performance of connectivity. In the future everything will be connected through embedded intelligence. Not only will people be communicating but things will also be connected. The first step toward the Internet of Things is to enhance the usability of the current devices by better connection to the services. All the capacity that can be provided is likely to be used. The hype about ubiquitous communication will not be realized unless extremely cheap, small, low-power wireless connection devices can be created for personal area and sensor networks. The goal is to have data rates which are comparable to wireline connections today. This will require very wide bandwidths with radio frequency (RF) operation frequencies in the gigahertz range.

However, a wider bandwidth is not the only way to obtain more capacity. As discussed later in this chapter, cognitive radio will provide a new paradigm to embed intelligence not only at the application level but also in the radio connectivity. In brief, it promises to share radio spectrum autonomously and more dynamically than previously, thus allowing better spatial spectrum efficiency, i.e., more capacity using the same bandwidth. Local area networks in particular would benefit from this approach because holistic radio resource management will become more complex in densely populated areas with several radio systems operating in the same frequency band. Also the concept of a cognitive network has been introduced to describe autonomous optimization of network performance. This allows the best available end-to-end data transfer in a complete network.

As already discussed in Chapter 4 signal processing will face challenges during the next ten years. The manufacturing of chips for computing and wireless requires huge

Nanotechnologies for Future Mobile Devices, eds. T. Ryhänen, M. A. Uusitalo, O. Ikkala, and A. Kärkkäinen. Published by Cambridge University Press. © Cambridge University Press 2010.

resources, thus the business has become focused on a few players. Here an alternative technology for mobile devices may be offered first to bridge the gap between the old and new technologies. For example, nanotechnology may enable thin, flexible, and tunable antennas.

Near field communication based on radio frequency identification (RFID) tags – probably in the future produced using printed electronics – is one of the first steps towards the smart space. The cost, size, and extremely low power consumption are the main drivers for the "low-end" connectivity for personal area networks, RFIDs, and other short-range communications.

In order to emphasize some opportunities for nanotechnologies in the field of radio communication this chapter will first focus on the principles of different communication concepts. Then we will address some of the fundamental requirements of radio communication including the scalability to different usage patterns. RF signal processing from the antenna to the digital conversion of bits is briefly reviewed as well as the hierarchy of the radio system design from materials to complete radio implementations. The remaining signal processing follows the same principles as given in Chapter 4. However, the signal processing requirements are imposed by capacity, reliability, and real-time requirements of the transferred data. The focus here will be on portable devices, and infrastructure, i.e., radios with fixed and constant power supplies are considered only as system elements. Considering the opportunities and challenges of nanotechnology in the field of radio communications will let us envision new concepts, as well as potential opportunities and disruptions that nanotechnology may offer in radio applications in the future.

6.2   Some principles of radio communications systems

The commercial use of radios for private consumers was popularized via broadcast radio transmissions. Later technology allowed television transmissions that further revolutionized media distribution from news to entertainment. However, widespread personal access to the limited radio resources required at least three major steps in the technology, i.e., adoption of digital transmissions, the cellular network concept, and affordable electronics enabled by extensive integration of functionality to silicon-based application-specific integrated circuits (ASICs). These allowed the transfer from early paging and analog telephone systems to the extensive digital communication of today.

The reliable information transfer capacity is fundamentally limited by Shannon's theorem [1]. This gives the theoretical relation between bandwidth, signal-to-noise ratio (SNR), and the largest achievable error-free transmission rate of digital data. Despite a continuous chain of innovations, the limit has not been achievable so far in practical systems. In addition, the need to mitigate interference (i.e., optimization of the total capacity) has led to regulated use of the spectrum over the technologically feasible range from the early days of radio communications. For example, current regulations set by national and international bodies determine the use of the spectrum up to 300 GHz. However, the most valuable part lies at the low end of the spectrum due to smaller

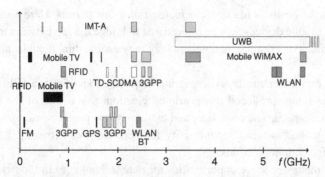

Figure 6.1 Some current and future radio bands used in portable consumer devices. Note that many of the frequency allocations are regional.

atmospheric attenuation of signals and lower implementation cost. Figure 6.1 shows some of the most important frequency bands and standards used or planned to be used in traditional mobile phones, laptop computers, and other kinds of multimedia devices. The heterogeneous structure arose from both technical and business aspects, as will be discussed next.

In business, exclusive ownership of the spectrum, like in cellular telephony, allows easier optimization of the capacity. At the same time, too rigid a spectrum allocation leads to an underutilized spectrum. However, building global coverage requires growing infrastructure investments from individual operators to meet the increasing data rate requirements. The availability of the wired Internet in the built environment has simultaneously given new opportunities to build local wireless Internet access with reduced coverage using less strictly regulated industrial, scientific, medical (ISM) bands at a reasonable cost. This has lead to a situation where several different radio technologies can provide the same services for consumers with acceptable user experience.

The key technical constraints in all radio systems, applying to both individual links and the network, include:

- operation range of a single link and coverage of the whole network;
- interference management;
- data rate and latency of the service;
- mobility of the devices (speed and seamless handovers of connections in the network);
- transmitted power and total power consumption (especially in mobile devices);
- time, frequency, and spatial domain behavior (e.g., precise moments of transmitting data packets at certain carrier frequencies).

These can be mapped to the radio protocol requirements and to the physical specifications that a single radio device must meet. Spectral efficiency and system capacity are the prime targets in high-speed radio access systems. Several new technologies have been and will be adopted to circumvent earlier bottlenecks. These include multicarrier modulation schemes, multiantenna solutions like multiple-input and multiple-output (MIMO), dynamic spectrum allocation, and cooperative communication between devices.

In this chapter, we will emphasize some of the details that have a specific impact on the opportunities to adopt nanotechnology or any other new technology in radio implementations. One must bear in mind that the different standards and operational principles are based on the best estimated technical performance for one or several use scenarios at the time of the standardization process. Evolved versions or new standards have utilized new technologies but so far no existing standard has been able to meet all the future needs. On the other hand, the new standards may not provide major financial benefits over the prior investments and thus will not allow direct replacement of the existing services in the market. Digital radio broadcasting, for example, has not replaced frequency modulation (FM) radio. However, many of the radio channels are also available in the Internet. Hence, they are wirelessly available for consumers over different radio protocols. This gives the freedom to choose and the possibility to get services from the easiest source. Unfortunately, this may lead to an unintentional inefficient use of radio resources and consequently battery waste and increased interference. By using this simple example, we can reveal the complexity of the converged world from the radio communications perspective.

6.3 Wireless communications concepts

From the radio implementation perspective, broadcasts (radio and mobile television) and satellite navigation systems (like global positioning system (GPS)) establish one distinct category although data rates may be very different. In a mobile device for such systems, only the radio receiver is required. Due to the long link distance and broad coverage, the received signals are typically close to the thermal noise level, and tolerance of strong interferers and fading is required in the mobile environment. The interfering signal does not necessarily come from an external source but from a transmission from another radio inside the same device. One specific example of new features for old standards is the very low-power FM radio transmitter in portable devices. The solution can transfer audio content (e.g. decoded MP3 files) from a small device using a wireless link to car radio systems.

Cellular communication systems have been the main drivers for wireless industry growth since the early 1990s. Second-generation digital networks are optimized for speech, whereas the third generation is optimized for high-speed data. This evolution has led to requirements for increased data rates and spectral efficiency. On the other hand, coverage of a wide area requires large maximum transmission power but also effective power control techniques to minimize radio interference and power consumption especially in mobile devices. The core principles include centralized control within the network, smooth handovers of transmissions over the cell boundaries, and designated bands for different operators. Developments in the field have brought Shannon's limit closer to reality than ever in LTE (long term evolution of 3GPP) which will be deployed commercially within the next few years. The future trend in cellular communications also will be towards improved seamless local access.

The other major development path has been the wireless local access network (WLAN) that was originally developed for wireless Internet access for portable computers. However, its use in other portable multimedia devices is quickly increasing. Because WLAN uses modestly regulated ISM bands, the systems can be deployed for many different purposes from private homes to public city area networks at a reasonable cost. WLAN radio access is based on collision avoidance schemes instead of centralized control, and different networks can share the same frequency band up to a limit. Coverage and power and spectral efficiencies are lower than with cellular systems. This is due to different regulatory aspects, and also to the less limited power source of the original application (i.e., Internet access using laptops) in the system definition phase. In addition to the scheme with fixed access points as local stationary control elements, WLAN can also be operated in the ad hoc mode where point-to-point connections are possible between mobile devices.

However, power consumption has so far been a major bottleneck for small battery operated devices in the ad hoc mode. Thus, Bluetooth (BT), as a less power-consuming system, has been widely adopted for mobile use, e.g., in wireless headsets and moderate-rate data synchronization tasks. However, the maximum data rates of WLAN cannot currently support very high-speed applications like rapid transfer of large files or high-definition television (HDTV) without cables. To enhance the maximum capacity locally, ultrawideband (UWB) systems operating under the 10 GHz range have been developed. A more recently proposed alternative is to use the 60 GHz band where several gigahertz of less crowded spectrum are available. The increased speed of nanometer-scale complementary metal-oxide semiconductor (CMOS) technologies has brought the band close to commercial adoption for mass markets.

Mobile ad hoc networks (MANETs) are attracting increasing research interest due to their self-configuration capabilities to establish local communities that evolve dynamically. In principle, these can use any point-to-point radio link, like ad hoc WLAN, but routing and packet forwarding of data from other users potentially increase the activity (i.e., radio time per device) and hence the power consumption of individual radios. Thus, power efficiency both in local communications as well as in processing is a critical element in the mobile nodes of the system.

RFID and radios for small sensing devices are examples of communications where typically a small number of local data are transferred to a reading device. These approaches require extremely low power consumption and a low price for the radio that is attached to the remote node. The desired service-free lifetimes of the remote sensors need to be in months or preferably in years with the cheapest batteries. The power consumption of such devices is typically dominated by the radio. The active power consumption needs to be very carefully traded off with the range (from centimeters up to 10 meters), but in addition extremely power-efficient protocols that minimize the duty cycles of individual radios are required. This applies to both system architectures, where sensors are supplying the information to central reading stations or fusion centers via mesh networks, and sensors that are infrequently read with mobile reading devices. This has led to a large variety of customized, proprietary radio solutions optimized for specific applications.

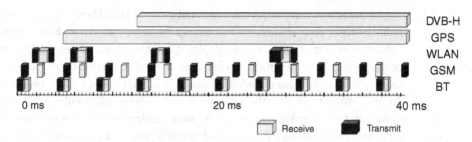

Figure 6.2 One possible scenario of radio activity in a multistandard radio device where multiple receptions and transmissions of different systems overlap in time causing mutual interference in the device.

The BT low energy standard was initiated to bring a uniform approach for the mobile device ecosystem.

RFID systems supply an equally scattered market that covers a large range of applications like product tracking and supply chain management, electronic payment, and electronic road tolls. Reading very cheap, passive tags using mobile devices is a fundamental challenge due to the remote powering of the tag. On the other hand, the cost structure of semipassive (battery-assisted) and active tags limits the potential applications. To complete the discussion on sensor-related radio communication concepts, we still address body area networks (BANs), where a lot of research emphasis is currently put on patient monitoring with different kind of sensors attached on or even under the skin. Sensor aspects have been discussed in Chapter 5.

6.4 The state-of-the-art and some future trends

Modern mobile multimedia devices are probably the most complex radio devices in the history. The rapid pace of hardware miniaturization together with diverse demands of consumers for handheld devices cannot be satisfied using a single radio. A state-of-the-art multiradio device includes cellular access (including multimode 2G/3G operation and also 4G/LTE in the future), WLAN, BT, GPS, mobile television (DVB-H), and FM radio. In a multistandard radio device, many usage scenarios require multiple radios to be active at the same time, i.e., they are in practice sending and receiving data packets simultaneously in different frequency bands, as illustrated in Figure 6.2. Due to different radio standards, different operation principles, and unsynchronized networks (if any), some of the radio transmissions inevitably overlap in time. This means that transmitters inside a single device can be the dominant sources of interference in receiving systems even though they are not occupying the same frequency band. Mutual coupling between components, and especially antennas, brings a part of the transmitted power to the received signal path. Then, nonlinear transistors and other active components in the transceiver transpose the interference over the desired signal. Alternatively, out-of-band noise and harmonic distortions of the transmitted signal will leak directly on the band of interest.

Computational complexity and controllability inside and between already established radio systems pose challenges that cannot be solved solely with the improved performance of semiconductor technologies. Cognitive radio was originally proposed to enhance programming capabilities and more autonomous reasoning of software-defined radios (SDR) [2]. It also included the idea of having improved interoperability between different systems sharing the same spectrum with so-called radio etiquettes that can make spectrum regulations ultimately more moderate. This has inspired research to seek new more intelligent ways to obtain efficient, more autonomous, and dynamic use of the radio spectrum [3]. Improved spectrum utilization is once again the current main driver of the research. Radio-scene analysis, i.e., searching for free and occupied channels even when buried under the thermal noise in the receiver, has been one of the main focus areas. Spectrum sensing is a key element to enable better autonomous decision making to launch communication at a certain frequency channel and time, or collaboratively share the spectrum. Cognitive techniques will be commercially applied, e.g., to establish local radio connections at sparsely occupied TV bands in rural areas. At the moment, many aspects of the cognitive radio are under extensive research and debate, and in the USA, the Federal Communications Commission (FCC) has already released rules to allow so-called secondary usage of unlicensed radio transmitters in the broadcast television spectrum. In many cases cognitive use of spectrum would require revised radio regulations by the International Telecommunication Union (ITU). These will be discussed in the future World Radiocommunication Conferences (WRC). Naturally, increased computational load will also result in significant technological challenges especially in mobile communications in the future.

As already discussed, the wireless technology field is currently rather fragmented, but the opportunity for scalable and converged services for consumers is provided with multiradio devices. Cellular access, WLANs, and broadcast systems guarantee continuous connections to information almost anywhere via the Internet and other sources. In many applications, the quality of service (QoS) scales with the data rate of the connection. However, the rapidly increased Internet traffic, especially in hot spot areas, has required the development of new generations of wireless systems. For example, the rate of cellular data has increased from about 10 kbps in the original GSM to several Mbps with the third-generation high-speed packet access (HSPA), and will increase to tens of Mbps of average throughput per user with LTE in the coming years. A similar evolution has also taken place with WLAN and many other standards.

Mobile Internet devices will operate as elements of the wireless grid that provides multiple ways to access the Internet and process local and global information. One can consider this as a wireless overlay that contains local context-dependent information including communications, as well as a multitude of means to connect to the Internet, as illustrated in Figure 6.3. One future scenario is that nanostructures will be connected to local grids and hence to the Internet via a multitude of local arbitrators that amplify and forward the weak signals from the nanostructures. An arbitrator can be a more powerful radio attached directly to the surface of nano islands. These islands will have local means of internal communications. The arbitrator can also be a nomadic, even transformable, mobile device with a larger energy source, or a low-cost accessory powered from an

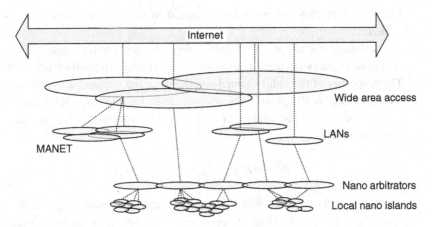

Figure 6.3 Heterogeneous wireless overlay including scalable islands of local processing and connections to the internet.

electrical socket, or a sensor node with a solar panel or some other energy-harvesting method. A dense grid with distributed ultralow-power processing elements may be technologically the most sensible choice for nanotechnology communications due to the scale of the physical quantities. The extremely localized processing elements, or even surfaces, will need specialized interfaces. The interface can be either a structure with an immense number of parallel nanopower radios or a conventional radio device with a "radio-to-nano" interface that amplifies the transceived information from nano- to micro-scale. Finally, the ultimate density of the grid will depend on the capability of fabricating extremely low-cost units of information-processing elements with a communication capability.

Because the local processing elements probably will not be able to create global or even locally comprehensive information space, interfacing to the rest of the world will require more mature radio technologies. Another much more conventional approach for nanotechnology is just to replace some of the existing radio components with new nanoscale devices. However, interfacing between technologies of different form-factors will result in a new set of challenges. The rest of this chapter discusses key technical aspects related to radio implementations and addresses opportunities and boundaries related to nanotechnology.

6.5   Radio system requirements

Extensive radio modeling and analysis especially on the network level requires a diverse set of parameters, a lot of computational power, and finally field testing for system optimization. Here we will focus only on some basic principles related to individual radio links and give an overview of some of the key parameters of the existing systems. A set of different kinds of radios is selected to illustrate different operational concepts rather than giving a comprehensive presentation of all options available in the market.

Magnitudes of different parameters, instead of detailed specifications, are given in Table 6.1.

The radio performance at the link level depends simply on the range, radio propagation environment, and susceptibility to noise and interference from other radio transmissions. The properties of the radio channel together with the required data rate and quality of the signal (i.e., SNR) provide the ultimate limits for implementation. The link budget equation gives a simple notation of the key performance parameters between a receiver and a transmitter:

$$P_{RX} = P_{TX} - L_{path} - L_{fade} + G_{TX} + G_{RX}, \qquad (6.1)$$

where P_{RX} is the received power, P_{TX} is the transmitted power, L_{path} is the path (free-space) loss of the radio channel dependent on the length of the radio link, L_{fade} is the fading margin dependent on the properties of the radio channel, and G_{TX} and G_{RX} are the antenna gains (i.e., directivity and efficiency) of the transmit and receive antennas, respectively. In (6.1), all the quantities are given in the decibel scale. Equation (6.1) indicates that the theoretical maximum line-of-sight range of the link is typically much longer than that which is achievable in practical conditions taking multipath-induced fading, blocking, and other effects into account.

The minimum received power (i.e., the sensitivity of the receiver) is determined by the thermal noise within the received bandwidth, the internal noise (figure) of the receiver, and the minimum SNR for achieving a certain bit error rate (BER) that allows decoding of the digital transmission. Also the receiver structure and detection algorithms play an important role when the sensitivity of the receiver is determined. In different radio standards, typical noise level requirements are roughly 8–20 dB above the thermal noise floor depending on the application scenario. Short-range (<100 m) radio links have more relaxed specifications to achieve more tolerance to interferers, simpler architecture, and lower power consumption. However, range is important in all radio systems, and therefore sensitivity is one of the key competitive parameters. With current technologies, many implementations targeted on mobile devices achieve noise figures in the range of 5–10 dB measured at the antenna connector in typical conditions. Some decibels of margin need to be reserved for static (such as manufacturing) and dynamic (such as temperature) parameter fluctuations in different components.

In general, higher bit rates require wider bandwidths for transmission, or more spectrally efficient modulations demanding higher SNRs, or multiantenna transceivers with MIMO. Hence, more transmitted power is required at higher data rates if we assume a constant noise figure of the receiver and a stable radio path. In practice, wider bandwidths and higher-order modulations complicate the transmitter design by increasing either power consumption or interference at adjacent frequency channels. Finally, the free-space loss of the radio channel increases as a function of frequency, which is an issue at high radio carrier frequencies where there is potentially more bandwidth available. Thus, the link ranges at higher data rates will be shorter in practical applications with the exception of fixed infrastructure radio links using highly directive, large gain antennas.

Table 6.1. Examples of key parameters in different kinds of radio systems

	Cellular 2G GSM/EDGE	Cellular 3G WCDMA/ HSPA	Cellular 4G LTE	WLAN 802.11a/g	WLAN 802.11n	BT	Sensor[a]
Typical range[b]	<35 km	<5 km	5–100 km	40–150 m	70–250 m	10–100 m	10–100 m
Signal bandwidth (MHz)	0.2	5	1.4–20	20	20/40	1	vary
Max. data rate (Mbps)	up to 0.24	up to 14.4	up to 350	up to 54	up to 600	1 … 3	0.01 … 1
Operating frequencies (GHz)	0.38 … 2.0	0.7 … 2.7[c]	0.7 … 2.7	2.4/5.8	2.4	2.4	0.4/0.9/2.4
TX max. output power (W)[d]	0.4 … 2 W	250 mW	200 mW	100 mW	40 mW … 1W[e]	1/2.5/ 100 mW	0.25/1/ 100 mW
TX min. output power (W)	1 mW	10 nW	100 nW	N/A	N/A	1 µW	N/A
Sensitivity (W)[f]	63 fW	2 fW	100 … 790 fW[g]	320 pW	6 … 790 pW	100 pW	vary
RX max. signal (W)	32 µW	3 µW	3 µW	1 µW	1 … 10 µW	10 µW	10 µW

[a] Some typical examples because many standardized and proprietary protocols exist.

[b] In 4G, the longest ranges require transmission via a link.

[c] Multitude of bands at different frequencies over the given range.

[d] Some typical numbers, in most cases specifications allow different power classes (i.e., maximum transmitted output power of a specific device).

[e] Radio regulatory classes vary in different regions and countries.

[f] Specified for one or more typical modulation and coding schemes in the system. In practise, significantly better sensitivity than specified is required in the implementation.

[g] Depends on the RF frequency band and signal bandwidth.

The range of a single link is limited not only by noise and other non-ideal effects at the band of interest but also by other transmissions in the same or adjacent frequency carriers or further away. To regulate the out-of-channel interference caused by nonlinear effects and noise in transmitters, all radio standards specify a transmission mask. For example, the unwanted power in the adjacent channel must be 30–60 dB below the maximum transmitted signal depending on the protocol. In radio receivers, the instantaneous dynamic range is determined by the selectivity, which means the capability to both filter out-of-band transmissions and avoid unwanted frequency conversions over the channel of interest caused by the nonlinear behavior in the receiver components. Once again, the magnitudes of the out-of-band signals depend on the offset frequency and the particular system. The largest blocking signals range from about 40 dB to almost 100 dB above the minimum received power in mobile applications. It is understandable that the dynamic range requirements in both transmitters and receivers are more demanding in wide-area systems like cellular communications than in local connectivity. Detailed descriptions of different nonidealities, transceiver system design and block level partitioning can be found in, e.g., [4].

In mobile applications, efficient power control is a key parameter to optimize overall system performance and power consumption. In radio transmitters, wide-area systems may require over 70 dB of output power control range to adjust the radio link power to the minimum required for reception with sufficient quality. In the 3G cellular systems, for example, the same frequency can be shared simultaneously with several users using code-division multiple access (CDMA) scheme. A fast power-control loop between the mobile device and the base station is required in order to minimize cochannel interference disturbing other transmissions at the same frequency carrier. On the other hand, short-range radios, like WLAN, may operate well without any transmit power-control scheme. In radio receivers, gain control is needed to level the desired signal at the antenna (with almost 100 dB dynamic range in some systems) roughly to a fixed value for digital signal decoding. This function, as well as filtering of out-of-band signals, is cascaded along the signal path, and it is partially done using analog and digital signal processing depending on the optimum performance of analog-to-digital conversion for the particular system.

In radio signal processing, power consumption depends strongly on the system specification and technology. In digital signal processing, technology has improved for decades as predicted by Moore's law, and area and power per function have been reduced. Unfortunately, the benefits are no longer that obvious in semiconductor technology nodes from 65 nm (equivalent gate width) downwards, taking the required investments into account. Also analog signal processing has benefited from faster transistors but not at the same scale. First, it allowed RF integration at frequencies of 1 GHz and above, and then improved performance with aggressive integration and architectural innovations. However, in the future, the profitability of scaling is also questionable for high-speed analog processing at the most used frequency range.

To conclude the observations of this section, Figure 6.4 gives a simplified view of the key tradeoffs in different kinds of radio systems. Power consumption is a consequence of link performance, meaning here the link length and dynamic range, and capacity including bandwidth. As the battery capacity in mobile devices strongly depends on the

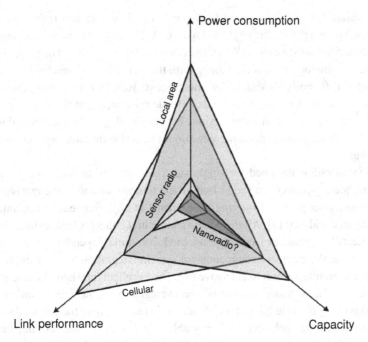

Figure 6.4 The trade-off between performance, capacity, and power.

form-factor of the device, the original application has a strong influence on the balance. For that reason, cellular access systems for mobile phones differ fundamentally from WLAN originally targeted on laptop computers. On the other hand, radios for sensor applications require a very long lifetime with a small battery, or some energy-harvesting solution. Hence, the performance must be scaled down to meet the power constraints. Based on the physical properties, one can envision that the first practical applications of nanotechnology will be in the sensor radio domain, perhaps with higher bandwidth and even shorter range. However, the feasible radio architectures are not obvious. The potential for distributed signal processing and locally improved computing power must follow the fundamentals of radio communications, but these will leave a lot of room for innovation.

6.6 Radio implementation architectures

The wireless signal processing functionality can be described using the open system interconnection (OSI) model targeted on communications and computer network protocols [5]. Only the lowest two layers (the physical layer and the data link layer) of seven have direct impact on the radio implementations if we neglect general-purpose computing tasks of central processing units (CPUs). The medium access control (MAC) sublayer of the data link layer (L2) provides the radio channel access allowing several radios to use the same medium. Hence, it defines, e.g., active states (i.e., the time-domain behavior) and used frequency carriers giving constraints on the implementation together with the physical layer (L1). The physical layer functionality can be roughly divided into two parts: RF and baseband (BB) processing. The reader is reminded that the terminology has evolved along with the technological development and is used somewhat differently depending on the context. Here, RF processing includes functions of air interface (i.e., the antenna), frequency conversion, filtering, amplification, signal level adjustment, and conversions between analog and digital domains. BB processing includes digital encoding, detection, channel estimation, synchronization, and decoding.

In digital radios designed for single standards, the RF architecture can use one or more frequency-conversion stages both in the transmitter and in the receiver including appropriately set gain and filtering components at each frequency to optimize system performance and cost [4]. All the basic architectural concepts have existed for decades, but different approaches have been adopted over time depending on the technology. For example, the extensive semiconductor system integration brought single-frequency conversion architectures into extensive use. The excellent transistor parameter matching of structures on the same piece of silicon overcame implementation bottlenecks of the solutions using discrete transistors. Hence, a cost benefit over the classical superheterodyne approach was achieved by being able to integrate all the filtering functionality (except for the processing at the actual radio frequency) into the same ASIC component. Another clear trend has been the gradual shift towards more digital and more programmable architectures. The digital signal processing performance has gained more from the reduced dimensions in ASIC lithography, consequently changing the optimum balance in the system partitioning. However, the possibility of direct digitalization at the antenna connector is limited by the signal conversion performance between analog and digital domains. The dynamic range of low-power analog-to-digital conversion in receivers, in particular, is improving slowly [6]. A more aggressive approach has been proposed for radio transmitters [7], but there also the power delivery capability directly from digital bits to the antenna is so far limited. In BB signal processing, a similar trend is taking place from customized circuits towards more general-purpose processing. This is limited by the very small power budget in mobile applications [8] because increased programmability inevitably causes some penalty in power consumption.

The complexity of radio implementation increases markedly if one allows several standards to operate or even share the same resources in a multiradio device. This applies to both control of functions and signal processing of the transferred data. Software-defined radio (SDR) has been considered to be a potential solution with different characteristics in RF [9] and BB [10] signal processing. Sharing at RF can mean that only a necessary number of parallel, programmable, 'general-purpose' RF signal processing paths will be configured depending on the number of active radio protocols. This is rather straightforward if the protocols are synchronized and transmissions are coming from the same operator. For example, 2G (GSM/EDGE) and 3G (WCDMA/HSPA) radio systems can share a lot of hardware because the handovers between systems are properly defined in the standard. That is not generally true, and the example illustrated in Figure 6.2 is much more challenging if the number of signal paths is smaller than the number of protocols using it. In the RF domain, signal processing must be carried out in real time because it is not feasible to buffer the signal to a memory. For that reason, separate signal processing paths are mostly needed in the currently available solutions and dynamic control is a challenging research question. The other major technology bottleneck in RF SDR is the tunability of high-frequency components, especially external filters. So far, an adequate performance (high selectivity and low loss) for any practical wireless system in mobile use could not have been achieved with a tuned center frequency. For that reason, parallel elements are needed to select RF bands. Also the tuning capabilities of antennas are still quite limited. In BB processing, some signal buffering is feasible giving more freedom to process several radio packets from different sources sequentially in the same processing unit. However, power consumption and cost (i.e., silicon area) may be acceptable only for the most complex multiradio solutions [10].

CMOS technology is dominant in the present single- or two-chip radio implementations with some external components like power amplifiers for the most demanding solutions, RF filters, and antennas. As nanotechnology is still to progress to mass fabrication, a straightforward comparison at the architecture level does not make sense. Attaining sufficient maturity to process circuits from multiple components is the indispensable next step, but we will still probably not be in a position to take advantage of most of the opportunities that can only be achieved when larger functional entities are integrated using the same technology. However, several quite different development paths are foreseeable that can address the issue from different perspectives in radio communications. Of the external components, antennas, high-accuracy reference clocks, and tunable filters could be candidates for faster adoption of nanotechnology. This is more likely than partly or completely replacing the highly integrated CMOS ASIC functionality. In computing, more power-efficient number crunching especially with machine learning could offer new options to manage complex real-time control of multiradio and allow opportunities for advanced cognition. All of these ideas are still a long way from being developed into practical solutions.

Another obvious need for computing is self-correction of local nonidealities. Many current RF solutions already contain digitally assisted analog elements to trim the performance to an acceptable level after all manufacturing and temperature variations. If one could make the border between analog and digital signal processing less explicit, more sophisticated control loops might be computed locally without loading the CPU. This would totally revolutionize the architecture in the long run and distribute processing even further across a radio grid or surface.

Because the electrical parameters in nanoscale devices are orders of magnitude different from those in main stream technologies, a distributed structure with a huge amount of parallelism in the system architecture level could be a practical platform for a "nanoradio." The first practical solution may be some simple communication system in the centimeter range, perhaps attached to a nanosensor implemented using the same technology. The information would probably be relayed with a transponder to longer distances.

The use of nanotechnology in complete radio systems is still in its infancy. Different approaches and technologies need to achieve a degree of maturity at each phase of the

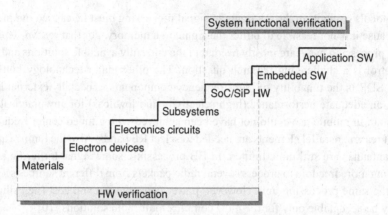

Figure 6.5 Hierarchical system architecture (HW – hardware; SW – software).

system hierarchy, as described by the staircase model in Figure 6.5. The model describes the SoC (system on chip) or SiP (system in package) flow but is generally applicable to complete mobile devices as well. Whether this will be a valid model for nanotechnology evolution as well remains to be seen. The research path can be accelerated by a proper alignment of different phases. Abstracted behavioral and functional models and suitable overlaps between the phases need to be estimated carefully in order to avoid misconceptions when different disciplines are required. Direct replacement of some existing components can speed up the development. However, it is likely that architectural or even conceptual innovations suitable for nanotechnologies will be needed to reshape the future of wireless communications.

6.7   Nanoelectronics for future radios

6.7.1   Introduction

Nanotechnology, with its methods for designing material properties, is displaying a totally new parameter space for electronic design, and this will inevitably change the current view of radio hardware in the long run. Nanotechnology will be utilized to enhance performance of the conventional components but its main influence will be in finding the novel applications, architecture, and working principles beneficial for the nanocomponents. However, not only are new technologies emerging, the existing ones are evolving, and for many future applications current semiconductor-based radio technologies will be adequate. It is therefore essential to recognize where nanoelectronics will truly make a difference and to understand the physical boundary conditions with which radio applications need to comply. This section discusses the benefits and exciting possibilities as well as problems and challenges of nanoscale electronics. It reflects the current understanding of the authors based on the properties of nanowires, graphene, and carbon nanotubes in particular.

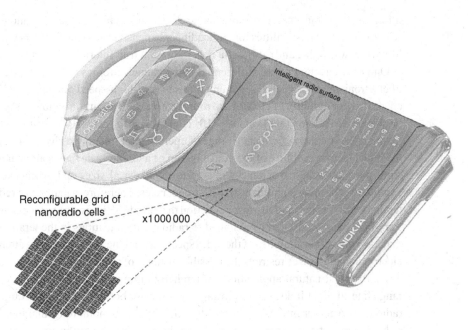

Reconfigurable grid of
nanoradio cells
x1 000 000

Figure 6.6 Intelligent radio surface with massively parallel nanoradio cells.

6.7.2   Potential of nanoelectronics in radios

The most obvious potential of nanoelectronics lies in the size and scalability it will offer
for RF architectures. While the digital electronics used in microprocessors and memories
has benefited from the shrinking of CMOS and Moore's law in the form of functionality-
versus-size, this has not happened on the same scale for analog RF electronics, because
traditional radio architectures rely on passive components like capacitors and inductors,
and unlike transistors their physical sizes do not scale down directly with diminishing
gate lengths. Nanoelectronics may bring a change to this with its new domain of operation
in which electrical quantities are no longer related to feature sizes in the same way as in
macroscopic devices, and massively parallel RF computing becomes feasible. Instead
of the highly optimized radio hardware for each standard and frequency band of today,
nanoradio of the future could have a universal RF processor in which different radio band
configurations are set from a vast grid of nanoradio cells, each dedicated for a certain
single frequency slot of the radio spectrum (Figure 6.6). When a different radio is
required, the system is reconfigured – or reprogrammed – accordingly. Redundancy will
be an integral part of this kind of concept as the number of available cells can be much
greater than is needed for any set of radios simultaneously. This gives new prospects for
hardware reliability as faulty cells can be automatically replaced by undamaged ones.

Ultimately, nanotechnology will change the physical form of radio equipment. Today's
two-dimensional, circuit-board-constrained form-factor of electronic hardware dictates
the appearance of the product. In the future, the surface of the mobile phone may act as
a part of its circuitry, containing a number of intelligent radios that adjust themselves to

different uses. Transparent and stretchable electronic circuits enabled by nanotechnology will give exciting possibilities for embedding inexpensive radio transceivers into surfaces like windows, walls, and fabrics – a necessity for a true Internet of Things.

One of the technological differentiators of nanoelectronics is the ballistic transport of electrons in nanostructures, which promises very-high-speed operation. In theory the charge carrier mobility of electrons at room temperature in carbon nanotubes (CNTs) can be as high as 10^5 cm^2/(V s) versus $\sim 10^3$ cm^2/(V s) for Si, or $\sim 10^4$ cm^2/(V s) for GaAs, giving rise to terahertz frequency range of operation for integrated electronics. However, the high electron mobility of the active region of a terahertz device is alone not enough, since parasitic stray capacitances related to the physical sizes and geometries of such components become significant at these frequencies. Therefore, a major size reduction of terahertz circuits is required – size reduction that is indeed the essence of nanoelectronics. Very high bandwidths can be assigned to radio communications in the terahertz region, thanks to the vast availability of the radio spectrum, but the high free-space attenuation of electromagnetic waves restricts the possible distance of communication at the same time. Thus, the most natural applications of terahertz radio communications are very short-range line-of-sight links, such as instant data transfer between handheld equipment and radio tags. Another interesting application of terahertz nanoradio is imaging. Terahertz radio waves act very much like visible light, but they have the ability to penetrate thin materials, heavy rain, and fog among things. It might also be possible to sense molecular resonances of harmful and hazardous gases with a miniaturized low-cost terahertz camera sensor. These features extend the possibilities of normal cameras and the human eye, giving a "sixth sense" for the users of terahertz imagers.

The discussed impact of nanoelectronics on radios is still just a prediction. Nano-technology is not mature enough for complete nanoradio implementation using the same technology platform throughout the whole system, and a lot of research topics have to be tackled before visions become reality. The first step towards nanoradios is to exploit existing radio architectures and circuit topologies but replacing the semicon-ductor components with nanoelectronic components. In fact, successful demonstrations have been reported [10] in which the transistor active areas are fabricated with parallel nanotube arrays. However, the resulting devices are still macroscopic. The risk of this approach is that since the performance of such a radio cannot compete today with a conventional semiconductor radio, which has been optimized close to perfection during the last decades, the potential of nanotechnology in this field is easily understated. As necessary as these demonstrations are for showing the current status of research, jump-ing to far-sighted conclusions based on them is not justified. Regarding nanoelectronics merely as an evolutionary step after CMOS, without changing the whole radio system paradigm, will not reveal the most fascinating opportunities it has to offer.

6.7.3   RF properties of nanowire structures

6.7.3.1   Impedance levels and matching

Due to their small dimensions, nanoscale devices operate according to different effec-tive models than macroscopic ones. Also the transport model depends on the regime.

Large-scale objects obey the diffuse transport mechanism, while Laplace equation and Ohm's law predict the performance of the system. For smaller objects the transport is quasi-ballistic when the size of the semiconductor channel is of the order of the length of the mean free path, the average length that the electrons can travel without collisions, (more than 100 nm for high-quality semiconductor heterostructures). The transport is no longer diffusive. It can be modeled using nonequilibrium Boltzman equations that give a statistical description of the state of the system. In the case of one order smaller (10 nm) objects the transport mechanism is ballistic and semiclassical, and is described by Sharvin's model, which states that the conductance is proportional to contact area. For even smaller objects (0.1–1 nm) in the ballistic regime, where the length of the channel is smaller than the mean free path, the quantum finite-size effects have also to be taken into account. These can be seen as oscillations in the conductance as a function of gate voltage. Landauer theory predicts that kind of system. In the case of ballistic transport the resistance is supposed to be zero. In this kind of system the contact resistance dominates the resistance. In theory the minimum resistance, i.e., the quantum resistance for a one-level (one-mode) wire is $R_Q = h/e^2 \approx 26$ kΩ, which is determined by the contacts. The actual resistance of a wire with a finite cross-section depends on the number of modes and thus on the properties of the wire such as the dispersion relation, which describes how the energy depends on the momentum of the electrons moving in the periodic lattice of atoms. For example, the minimum resistance of an n-type, two-dimensional semiconductor with a parabolic dispersion relation (typical of classical electrons moving at a velocity much lower than the velocity of light unlike photons or charge carriers in graphene) depends on the carrier density and is given by $R_{min} \approx (80 \, \Omega \, \mu m)/W$, where W is the width of the wire [12]. In contrast to the 50 Ω environment that has become the interfacing standard in RF electronics, or to the characteristic impedance of free space $Z_C \approx 377$ Ω in wireless information transfer, nanodevices have impedance levels of the order of kilohms. Practical impedance levels are even higher than that. Consequently, interfacing between nanoelectronics and the outer world suffers from the inherently large impedance magnitude difference. Without effective impedance transformation, it is practically impossible to transfer power and electrical waveforms across this impedance barrier without excessive loss.

In microelectronics, impedance matching techniques are used for maximizing power transfer efficiency in circuit interfaces, but their dimensioning becomes impractical when the impedance ratio approaches 1:10. This has proven to be one of the most difficult issues when considering the first foreseeable applications of nanotechnologies in conjunction with the existing radio hardware. It is a strong advocate of nanointegration, i.e., bringing as much radio functionality as possible to the nanoscale level with the least amount of interfacing to outside the nanodomain. Connections inside the nanointegrated system with a similar impedance environment can be managed more easily.

In addition to the quantum resistance, nanowires and nanotubes have also reactive properties like any other transmission lines. A suspended nanowire, illustrated in Figure 6.7, has in addition to the magnetic inductance, due to the varying magnetic field, and electrostatical capacitance, quantized capacitance (quantum capacitance) and inductance (kinetic inductance) values per unit length and per mode, arising from the

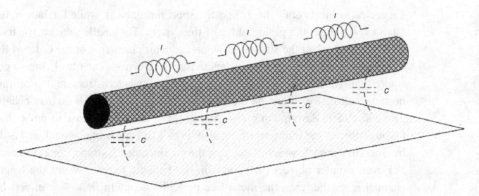

Figure 6.7 A nanowire as an RF transmission line.

quantum mechanical phenomena, i.e., from quantities such as the density of states, Fermi energy, and the kinetic energy of the electrons [13]:

$$c[\text{F/m}] = \frac{2e^2}{h\nu_\text{F}}, \quad l[\text{H/m}] = \frac{h}{2e^2\nu_\text{F}}, \tag{6.2}$$

where ν_F is the Fermi velocity of the material, e is elementary charge, and h is Planck's constant. For carbon nanotubes, $\nu_\text{F} = 800$ km/s, and thus the reactive elements become numerically $c \approx 400$ aF/μm and $l \approx 4$ nH/μm. The kinetic inductance is three orders of magnitude larger than the magnetic inductance. The characteristic impedance of a nanotube transmission line becomes

$$Z_\text{C} = \sqrt{\frac{l}{c}} = \frac{h}{2e^2} \approx 12.9 \text{ k}\Omega. \tag{6.3}$$

This is an important electrical property in wired nanotube interconnections and shows that the characteristic impedance is fixed and set by natural constants rather than the physical dimensions of the transmission line. Characteristic impedances of conventional transmission lines can be tailored by changing their conductor dimensions, and this technique is utilized for impedance transformation. With nanoscale transmission lines, this method fails. Instead, different impedance levels have to be synthesized with parallel nanotubes attached together. This results in quantized impedance steps and thus less freedom of design.

One clear consequence of the high impedance level of nanoelectronics is the inability to process RF power. While other potential applications of nanotechnologies – like computing – do not rely on certain signal power levels in their operation, the quality of a radio link over useful distances is very dependent on radiated RF power, as described earlier in this chapter. To produce power levels as low as a few milliwatts with the nanodevice impedance levels, a supply voltage of more than 10 V is needed, which is uncomfortably high for battery-operated devices and likely to be more than the destructive breakdown voltage of the nanostructure. It is clear that nanoeletronics has

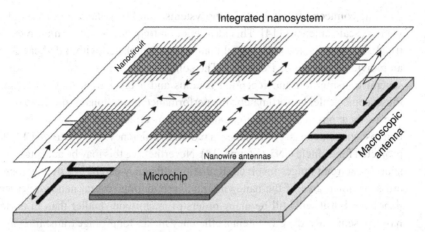

Figure 6.8 Wireless interconnections via nanowire antennas within a high-impedance integrated nanosystem and between the nanosystem and a low-impedance microsystem.

more potential for use in radio receivers, where there are less-demanding power domains, but even there power handling could become an issue in strong reception conditions.

As in the case of impedance matching, the high impedance level can be brought down by connecting a large number of nanodevices in parallel. This is a straightforward method of solving interfacing and power handling problems but at the price of size and functionality. It is difficult to control the integrity of a device with thousands of parallel nanostructures in which all of them should work more or less identically. Furthermore, driving such a device may require another low-impedance circuit in front, and the problem is just shifted one step back along the signal path.

6.7.3.2 Interconnections

As discussed in the previous section, interfacing between nano- and microscale circuits necessitates impedance matching if electrical information is to be transferred, and there are no good existing solutions to make the transformation. Inside nanoelectronic circuits, impedance matching is of less concern in the interconnections between different functional blocks and components. The superior current density capability of carbon-based nanomaterials (resistivity of around $1\,\mu\Omega$ cm vs. $1.7\,\mu\Omega$ cm of copper) makes them excellent conductors compared to, e.g., copper interconnects of the same dimensions. There is, however, a limit to how densely wired interconnections can be packed if currently available lithographical processes are used for fabrication. For highly integrated nanocircuits, this will be a bottleneck in achievable complexity and functionality, and necessitates new fabrication methods. One further issue to be considered is the contact resistance between two attached conductive nanoelements, as this is much higher than that which is seen of the macroscopic world, the order of $5{-}10\,\text{k}\Omega$ at best.

One of the most prominent propositions for circumventing the limitations of wired interconnections is wireless interfacing of nanosystems (Figure 6.8). Instead of galvanic connections, integrated nanosystems could be connected to the outside world via nanoscale antennas [14]. This idea relies on frequency-domain multiplexing of electrical signals in which each signal connection is replaced with a dedicated antenna of an individual resonance frequency. The concept would, in fact, create a radio cell with several transmission and reception stations and a local frequency spectrum, all at a mesoscopic scale. Unfortunately, the performance limitations of the nanowire antennas due to their physical dimensions and quantum effects are severe when compared to macroscopic antennas. For instance, a radiation efficiency of -90 dB or $10^{-7}\%$ has been projected for a single CNT antenna [14]. Nevertheless, this may be acceptable at the very short intracircuit ranges involved. Radiation efficiency can be boosted by increasing the cross-sectional area of the nanowire, i.e., by bundling several nanowires together as a nanofiber, but it will still be many orders of magnitude smaller than in normal antennas. Consequently, the low antenna efficiency makes long-range transmitter applications virtually impossible, as too much power would be wasted in compensating for it. One clear advantage of nanowire antennas is their inherent capability of impedance matching between low-impedance free space and high-impedance nanodomain, though with low efficiency. Hence, wireless interfacing may well become the only viable solution to the impedance matching dilemma.

6.7.3.3   High-frequency projections

The high resistance of nanoscale electronic devices does not limit their high-frequency performance in principle, since the inherent capacitance values are also very small, of the order of attofarads. Thus, the RC time constants become small enough for potentially even terahertz operation. For active elements like transistors, it is more relevant to look into the cut-off frequency f_T that indicates the highest frequency at which the device gain has dropped to unity. With only the intrinsic parameters of a nanotransistor considered, the cutoff frequency [15] is

$$f_T = \frac{1}{2\pi} \frac{g_m}{C_{gs}} \leq \frac{v_f}{2\pi L_g} \approx \left. \frac{130(\text{GHz})}{L_g(\mu\text{m})} \right|_{\text{CNT}} \tag{6.4}$$

where transconductance g_m is independent of the device dimensions but gate-source capacitance C_{gs} scales with the gate length L_g (Figure 6.9). The theoretical upper limit of the transistor speed is set by the Fermi velocity of the carriers, showing that terahertz operation is theoretically also possible for active nanodevices [15]. This opens new perspectives for terahertz radio integration, for which the technology does not exist today.

In current lithographically processed nanodevices, the optimistic predictions of high-frequency operation are somewhat diluted by the fact that parasitic effects are dominant in their performance. The sizes of electrodes (drain, source, and gate) are dictated by the device geometry inherited from the microelectronics world. Thus, their parasitic capacitances are significant when compared to the minute intrinsic capacitances of the nanostructures: they are practically of the same order of magnitude. The result is that the cutoff frequencies fall close to 10 GHz [16], which is not competitive with existing technologies. In fact, the parasitic capacitance issue is already apparent in

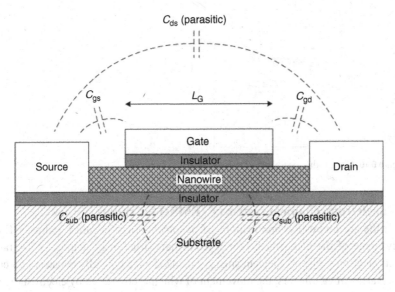

Figure 6.9 Nanotransistor with parasitic capacitances.

deep-submicrometer CMOS processes, but with the distinction that these can often be embedded in the surrounding circuit. It is clear that nanoelectronics achieve a break-through by exploiting traditional device geometries and using them as direct substitutes for semiconductor devices, as there is no justification in performance and most probably not in cost either. Instead, the whole device paradigm has to be brought into the nano-domain where all the parasitic restrictions are tied to the same physical environment as the actual functional phenomenon. On the other hand, compromised performance is acceptable in applications such as printable electronics, where silicon-based technologies cannot be used at all.

It has been suggested that ballistic (i.e., collisionless electron motion) nanotransistors could have extremely good noise properties due to shot noise suppression [17]. This would enable ultralow-noise amplifiers, and thus increased sensitivity of radio receivers, especially at operating frequencies beyond CMOS capabilities. The ultimate noise limit of such a low-noise amplifier is set by the quantum limit of noise temperature T_n in a high-gain linear system [18]:

$$T_n \geq \frac{hf}{k} \Rightarrow F \geq 1 + \frac{hf}{kT}, \tag{6.5}$$

where F is the noise factor of the amplifier, f is the operating frequency, and T is the ambient temperature. For example, a 100 GHz amplifier at room temperature has a quantum-limited noise factor of $F \approx 0.07$ dB, a figure that is not achievable with conventional techniques. This is naturally a very optimistic number for nanotransistors as well. Nevertheless, combinations of nanoscale ultralow-noise amplifiers and antennas will give rise to distributed radio surfaces with an immense number of individual radio receivers, giving totally new opportunities for physical form-factors.

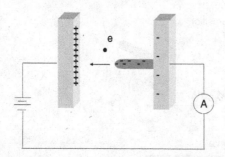

Figure 6.10 CNT radio.

6.7.3.4   Nanoeletromechanical systems (NEMSs) for radios

NEMSs are interesting applications of nanotechnology in future radios. NEMSs take advantage of the mechanical properties of nanostructures, such as CNTs, rather than their electron transport mechanisms. A NEMS device typically consists of a cantilever, or an array of cantilevers, the movement of which is controlled by electrostatic or piezo-electric actuation. The cantilevers can be CNTs or carbon nanofibers clamped to contact electrodes. Compared to microelectromechanical systems (MEMSs), the dimensions of the movable parts are much smaller resulting in greatly reduced controlling voltages (up to 50 V for MEMS versus a couple of volts for NEMS). This makes them much more compatible with the battery voltages of handheld devices. Another advantage for radio applications is the mechanical resonance frequency range of NEMS structures, which lies in the RF domain (100 MHz–10 GHz) due to their low masses and high Young's moduli. This gives potential for applications where high-quality resonators are required: filters, oscillators, and frequency references (replacement of quartz crystals). Other components that can be fabricated with NEMS include RF switches, voltage-controlled capacitors, and power sensors. NEMS also opens interesting possibilities for nanoelectromechanical system integration, where complex systems of NEMS switches and resonators with adjustable interdevice coupling may show novel functionalities for RF signal processing, yet on a single microchip.

6.7.3.5   CNT radio

CNT radio is a good example of how nanotechnology can be used to change the working principle of a device, how new enablers may lead to novel applications. A radio circuit is a construct of quite a few electronic elements: antennas, tuners, demodulators, and amplifiers. Jensen *et al.* have shown that a single CNT can perform all those functions making a complete functional radio circuit [19].

Figure 6.10 depicts the circuit for CNT radio. Firstly, a conductive CNT is clamped to an electrode, one end of which functions as a nanosized antenna. A dc voltage is connected to the electrodes. The applied dc bias negatively charges the tip of the CNT. At the same time, the electromagnetic radiation impinging on the CNT excites the mechanical bending mode vibrations in the tube. The length of the tube governs the resonance frequency of the system. The bias voltage across the electrodes produces a constant field-emission current, which is modulated by the mechanical vibrations due to the radio field.

As the CNT is vibrating a voltage is applied between the tip and a nearby electrode. The field emission of electrons from the tip of the CNT depends in an exponential manner on the distance from the CNT tip – which varies as the CNT is vibrating.

The field emission effect can be used to tune the RF. A high field-emission current is utilized to eject the carbon atoms from the end of the CNT, altering the length of the CNT and thus the resonance frequency. Fine tuning can be done by using tension due to electrostatic field.

The nanotube radio's sensitivity can be enhanced by reducing the temperature, using a lower resonance frequency, or improving the quality factor of the oscillating element. One possibility is to have an array of these components coupled to each other in order to have a higher resonance frequency and a larger quality factor.

6.7.4   Nanotechnology for radio front ends

In the context of this section, the term radio front end refers to the components between the antenna and the radio transceiver (including the power amplifier), i.e., essentially the RF filters and switches that are used to select the operating frequency band for each radio system (Figure 6.11). Front ends have been seen as the first radio applications in which nanotechnologies can potentially play a strong role. This view is amplified by the fact that current front ends cannot be integrated as no CMOS-compatible technology exists for fulfilling the stringent performance requirements. The front ends of today's radios are made of discrete devices laid out on modules, and thus are bulky and expensive. These problems are multiplied by the number of different radios in a modern mobile phone. Moreover, the fixed center frequencies and bandwidths of the front end filters will be major bottlenecks in flexible-spectrum-usage concepts including cognitive radio.

As discussed earlier, NEMS technology has been considered for future RF filter realizations. An RF filter is essentially a bank of coupled resonators, and so the individual resonator characteristics are responsible for the performance of the whole filter. Nano-electromechanical resonators have the potential for offering high quality factors at RFs at a small size, and their resonance frequencies can be altered via external control. All this sounds very attractive when considering the weaknesses of the technologies that are available today: fixed ceramic resonator and surface/bulk acoustic wave (SAW/BAW) filters of millimeter-scale sizes. Before looking deeper into the NEMS filter structures, it is essential to reflect on the performance requirements of front end filters.

6.7.4.1   Performance requirements of front end components

The RF front end has a key role in radio performance. It is responsible for delivering the weak reception signal from the antenna to the low-noise amplifier without adding noise and further weakening the signal. It also has to take the high-power output signal of the power amplifier to the antenna without dissipating an excess amount of RF power and distorting the transmission signal. These two functions must not interfere with each other, i.e., the strong transmission signal must not leak into the receiver side and block

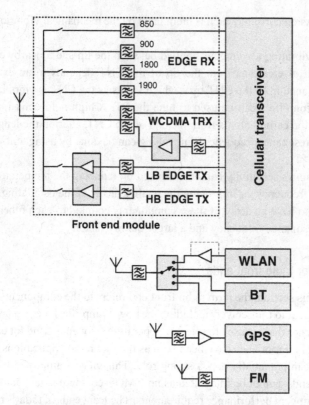

Figure 6.11 RF front end of a multiradio cellular phone.

reception of the same or other radio systems. The front end filters need to have large attenuations outside their own frequency bands in order to prevent this interference from taking place. Some typical numerical values of these requirements are listed below (see also Table 6.1):

- centre frequencies from 450 MHz to 5.8 GHz according to the radio standards;
- bandwidths ranging from 25 MHz up to 600 MHz;
- minimized insertion loss (typically <3 dB is achieved today, depending on the frequency band and system requirements);
- rejection of the receiver band (on the transmit side) >20 dB for time-division duplexing systems like a global system for mobile communication (GSM) (no concurrent transmission and reception) or >50 dB for frequency-division duplexing systems like wideband code-division multiple access (WCDMA) (concurrent transmission and reception);
- power handling capability of up to a couple of watts (on the transmit side);
- nominal impedance level of 50 Ω.

The specifications for insertion loss and out-of-band rejection translate into very high quality factors for the filter resonators, typically more than 1000, which limits the available resonator technologies to the few mentioned. It is notable that radio systems have been designed to take advantage of the good performance of fixed high-quality resonator filters, and therefore there is little flexibility left in the demanded parameters. This is bad news for any challenging front end technology as it has to be brought to perfection before it can meet the specifications. Hence, future radio systems with flexible spectrum usage should be designed keeping in mind the performance tradeoffs that adaptive front ends are bound to have. This will give more freedom choosing front end technologies and balancing their performance with the rest of the radio. On the other hand, short-range radios have much more relaxed front end specifications than cellular radios, and therefore they are likely to be the first applications of new integrated front end technologies.

6.7.4.2   NEMS resonators for RF filters

Applying nanotechnology to RF signal processing will enable above 1 GHz operational frequency. Obviously high frequency is related to short wavelength and thereby to small dimensions of the component. According to the classical theory of the vibrations of elastic plates, the high resonance frequency of the double clamped cantilever electro-mechanical resonator requires large stiffness (large Young's modulus), large width, small mass density, and a short cantilever.

CNTs are good candidates for NEMS resonators. CNTs are metallic, semimetallic, or semiconductive depending on the chirality, i.e., the spatial mirror symmetry properties of the tube arising from the different ways it is possible to roll up a graphene sheet to form a nanotube. For semiconductor devices it is possible to burn off the metallic tubes using high voltages, or the separation can be done chemically using material that is attached to the semiconducting tubes but not metallic ones. We can also use electrophoresis or dielectrophoresis by coupling the charge or the dipole moment (in the case of non-charged systems) to an electric field. The simplest way is just to control the growth of the nanotubes.

CNTs have fewer defects and better mechanical properties than nanowires (Young's modulus of the order of 1 TPa, large tensile strength, and large bending flexibility) [20], but the contacting of, e.g., silicon nanowires is more straightforward than integrating CNTs in the device. We could consider developing a better MEMS and CMOS compatible process, but integration would still be challenging. It would be better to build the whole system at the nanoscale instead of having connected MEMS and nanoscale components, or at least to build a bigger system at nanoscale. In order to build a real device we have to treat the interfaces between the nanoscale, the mesoscale, and the macroscale, and most probably this will cost energy. One of the main research questions is how to do this.

One solution is to build a resonator array or system of coupled resonators instead of a singe resonator in order to treat the large impedance. Transport properties of a system of 10 000 noncoupled nanotubes can be described as a conventional transmission line. Davis *et al.* have suggested building a nanotube array RF filter. They analyzed a nanoelectromechanical signal processing device based on arrays of CNTs embedded in RF waveguides [21]. The next step is to exploit the coupling of the resonators, which

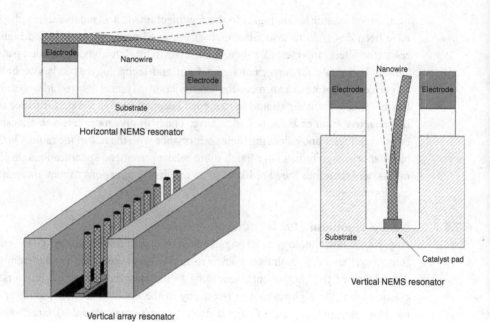

Figure 6.12 Different NEMS resonator configurations: a horizontal, a vertical, and a vertical array resonator.

may be useful for the device purposes. Treating a large nonlinear resonator array system stochastically we might be able to enhance the error tolerance, and thus reduce the challenge of the contacts.

NEMS resonator structures with CNT cantilevers are illustrated in Figure 6.12 [22, 23]. The cantilevers are laid either horizontally or vertically in respect to the substrate material. Horizontal resonator nanotubes are clamped at one or both ends to the electrodes. In the experimental designs, the nanotubes were typically grown elsewhere and positioned on the electrodes by electrophoresis. The fabrication of vertical resonator nanotubes is more controlled as they are grown in situ on top of catalyst pads. Theoretical studies have shown that these kinds of structures are able to resonate at frequencies of several gigahertz with substantial tuning ranges and high quality factors. However, the experimental results have not yet met the high expectations. Fabrication techniques for CNT resonators are still immature and have poor repeatability, and variations between samples are large. Electrical measurements are also inaccurate for single-tube resonators which are those most widely reported, because of the extremely high inherent impedance levels. Measured quality factors are still far from adequate, being less than 100, and the highest empirical resonance frequencies are only around 1 GHz [24].

For practical applications in radio front ends, and also for more relevant characterization measurements, much lower impedances of nanoresonators will be mandatory. This can be achieved with resonator arrays. Controlled nanotube growth is essential for attaining consistent geometries of hundreds or thousands of parallel nanotubes, and therefore vertical arrays are the most prominent. Although the size of one nanotube is very small, the array can occupy a relatively large area. The array elements cannot be placed in grids that are too dense, as nanotubes have a tendency to stick to neighboring nanotubes or electrodes due to their large surface forces. Thus, it is important to consider the tradeoff between the number of individual resonators, i.e., the impedance level, and the size of the composite geometry.

Other materials like silicon or ZnO nanowires and graphene nanoribbons are also being studied. These may well prove to be easier to fabricate than CNT resonators, as lithographical methods for patterning can be applied. It is also possible that nontunable nanoresonators will offer better quality factors and resonance frequencies than tunable ones. Obviously, they can be used as fixed-frequency and fixed-bandwidth filters that need to be connected as filter banks with a set of different center frequencies. Desired outputs are then obtained by reconfigurable inputs and outputs. This kind of structure is an integral part of the intelligent radio surface concept illustrated in Figure 6.6.

It is clear that NEMS resonator performance is not yet adequate to challenge traditional technologies, in spite of its theoretical promise. Looking at the radio standards of today, this will be true in the next few years as well. It is important to benchmark new technologies against existing ones, and also to understand their weaknesses. Nanotechnology will not be a short-term solution to radio front end integration. This is a strong argument for the statement that totally new concepts of radio and radio front ends have to be adopted in order to justify the use of nanotechnology. Whether the new concepts will follow the visions discussed in Section 6.7.2, remains to be seen.

6.7.4.3   Thin, flexible, low loss, tunable antenna

The size, efficiency, communication range, and cost of the wireless device are factors that limit the mobile connectivity between the user of the device and the rest of the world. RF circuits are difficult to miniaturize without compromising performance. Miniaturization of antennas is limited by the physics of electromagnetic radiation as the antenna is a transducer designed to transmit or receive electromagnetic waves i.e., antennas convert electromagnetic waves into electrical currents and vice versa.

Accelerated charge emits radiation. A transmitting antenna is a system of conductors that generate a radiating electromagnetic field in response to an applied alternating voltage and the associated alternating electric current. A receiver antenna, placed in an electromagnetic field, will induce an alternating current in the antenna and a voltage between its terminals.

Current technology uses bulk and thin-film forms of magnetic and conductive materials in the fabrication of antennas. As antenna sizes become smaller, losses due to electromagnetic dissipation in materials increase and directly affect the performance by reducing the gain. Also, electromagnetic interference (EMI) is a commonly encountered problem in modern mobile devices with multiradio systems. In addition to the desired electromagnetic field any conductor with a varying current radiates, and so acts an unwanted antenna.

The main stream of research and development has been to design a small, smart antenna with defined impedance bandwidth characteristics and improved directivity of the signal, e.g., by optimizing the geometry of the antenna, i.e., the current path on the antenna element, or choosing the materials. The design of the antenna depends on the design of the whole device. This makes the development work time consuming, but numerical simulations have enabled us to predict the performance at an early phase of the production cycle. The size is still an important limiting factor, as is the design time. Using nanotechnology will give us more design space as it is possible not only to choose the material but also to design the material properties in order to optimize the antenna performance.

The ultimate limitations of the antenna size, bandwidth, and gain are determined from basic theoretical considerations and can influence the design of any portable system, e.g., bandwidth and antenna size are inversely related. The bandwidth of a "small antenna" can be increased by decreasing the efficiency, thereby reducing the gain. With the new technology we can design material for the substrate of the system in order to have a miniaturized antenna with the same resonant frequency. The electromagnetic properties of materials such as cobalt or iron can be designed, e.g., by increasing the surface area of the material by building magnetic nanoparticle polymer composites. The quality of the antenna remains the same when in parallel with decreasing the size of the system we also increase the permittivity or the permeability of the substrate material [25].

The ultimate goal of the design of the materials is to build one antenna for all purposes from new externally controllable and tunable electromagnetic materials. By changing the type, size, and number of magnetic nanoparticles it is possible to tailor the magnetic permeability and hysteretic loss, electric conductivity, permittivity, and its loss tangent properties. Multiscale simulations combined with tailored measuring methods (to treat the frequency range and the form-factor of the system) are important for designing the material because the basis of the effect lies in quantum mechanics and in the structure of the magnetic nanoparticles.

6.7.5   Graphene for ultrafast electronics

The properties of future devices can be enhanced by designing the material properties using nanotechnology-based functional materials, the properties of which can be controlled externally. These new materials can be added to CMOS technology, e.g., to design fast transistors, but a more powerful way of using nanotechnology is to study the exceptional properties of the material that arose from the nanoscale effects. One of the materials with the greatest potential for ultrafast electronics and new kinds of devices is graphene.

Graphene is a two-dimensional allotrope of carbon. It is strong, stiff, electrically and thermally conductive, and its electrical properties can be tuned, which makes it a good candidate for future radio applications. Graphene is a hexagonal two-dimensional lattice of carbon atoms, where three valence electrons of the carbon atoms are in sp^2 covalent bonds in a plane, and the fourth electrons are conduction electrons.

The basic solid state physics theory of graphene has been known since Wallace published his paper about the properties of graphene in 1947 [26]. The hexagonal lattice is not a Bravais lattice, i.e., a specific three-dimensional geometric arrangement of the atoms composing a crystal defined by a lattice vector, but it can be modeled as a combination of two triangular sublattices that are Bravais lattices – identical and shifted.

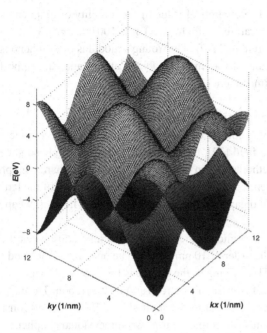

Figure 6.13 Two energy levels of graphene as a function of the wavevector k. The conduction band on the top and the valence band on the bottom meet at the Dirac points. The dispersion relation of graphene is linear in the vicinity of the Dirac point.

The symmetry of the system of two shifted identical sublattices with tight binding and interactions only with nearest neighbors leads to exceptional energy eigenvalues with a linear dispersion relation, which means that the energy is linearly proportional to the momentum of the electron moving in the periodic lattice of atoms. In graphene the electrons move at constant velocity which does not depend on their energy. The behavior is like that of light where increasing the energy reduces the wavelength but the velocity remains constant. In graphene charge carriers are relativistic Dirac fermions near the so-called Dirac points, i.e., the points where the conductance band and the valence band meet (see Figure 6.13).

It was thought that free standing two-dimensional lattices are thermodynamically unstable and do not exist, but in spite of this fact Novoselov and Geim found graphene layers in 2004 [27], and the existence of the graphene was finally confirmed by experimental observation of the quantum Hall effect, an effect that describes how conductivity is quantized on the magnetic field and that was theoretically predicted for two-dimensional graphene [28]. Because of the instability it is difficult, or impossible, to grow graphene sheet without a substrate. It seems that the substrate stabilizes it as it also opens the band gap. Crumpling in three dimensions may also stabilize a two-dimensional sheet. According to theory and experiments it opens the band gap.

Graphene is the best conductor at room temperature known. The measured mobility of the charge carriers at room temperature is 20 times the best silicon MOSFET, 20 000 cm^2/(V s) [27]. Its carrier density is more than 10^8 A/cm^2, and its mean free path is approximately 1 μm [29]. The theoretical value of the mobility of graphene is 200 000 cm^2/(V s) at the room temperature, while for CNT it is 100 000 cm^2/(V s).

Graphene is also transparent and stiff. The Young's modulus of graphene is approximately 1 TPa. It is chemically stable, impermeable to atoms, and can support pressure differences larger than 1 atmosphere [30].

6.7.5.1   Tuning the electrical properties of graphene

The electrical properties of graphene can be tuned using several techniques: nanoribbons [31], external electric field [32], the substrate below the graphene sheet [33], or crumpling. Using these methods we can open the band gap, i.e., change graphene from a metallic conductor to a semiconductor by breaking the symmetry of the lattice. Spin–orbit coupling opens the band gap, too, but the effect is small. The energy gap is smaller than 0.01 K [34].

Graphene nanoribbons are semiconductors at room temperature when the width of the graphene ribbon is of the order of 10 nm or less. The opening of the band gap is due to same effect as in CNTs. Cutting of the graphene layer to form a ribbon introduces a quantization due to the confinement of electrons around the ribbon. The corresponding component of the wavevector parallel to the short side of the ribbon is constrained by the boundary condition. Thus, in going from a two-dimensional graphene sheet to a one-dimensional ribbon, each graphene band is split into a number of subbands, with allowed energy states. Those cases with bands that pass through a Dirac K-point of the graphene Brillouin zone result in a metallic ribbon; all other cases yield semiconducting ribbon. The energy band gap is inversely proportional to the width of the ribbon [31], thus it is possible to tune the electrical properties of the graphene nanoribbon by changing the width of the graphene sheet and building systems of different geometries. This band gap engineering has been suggested to be the basis for a new kind of electronics. According to the theory these structures are sensitive to edge effects. Most interesting are the ribbons with zig-zag edges as these ribbons are magnetic, which makes them possible candidates for spintronics. By using different edge geometries and an external electric field it is possible to control the spin of the charge carriers at different parts of the transducer.

Graphene on a substrate or under the influence of a periodic potential [35] will break the symmetry of the graphene lattice and will open the band gap. In pristine graphene the group velocity of the charge carriers is theoretically 10^6 m/s. By changing the periodicity or amplitude of the potential, it is in theory possible to change the group velocity of charge carriers at different angles. The gate structure for nanoscale periodic potential requires something more than the standard top-down lithography that currently has limited resolution, but these methods are being developed. New inventions in the area of nanotechnology; e.g., superlenses based on metamaterials will make it possible to process nanoscale structures using top-down lithography. The method of self-assembly with block copolymers or DNA origami can be used to enhance the resolution of the conventional lithography techniques. Using conventional methods it is straightforward to create structures with dimensions of 100 nm. Depositing the polymers on the structure, e.g., a matrix of holes, and transforming the phase of the composite under specific conditions so that the competing polymers pattern to nanoscale structure, it is possible to multiply the density of the nanoelements [36].

The bandgap is opened for graphene layers that are epitaxially grown on a silicon carbide, SiC, substrate. The opening has been measured to be the largest on the single graphene layer near the substrate (a gap of 0.26 eV) and decreases as the thickness of the graphene layer system increases. It vanishes in the case of four layers [33]. The first layer on a SiC surface acts as a buffer layer and allows the next graphene layer to behave electronically like an isolated graphene sheet [37]. This interaction depends on the interface geometry.

Another graphene layer can be used as a top gate or like a substrate on a graphene sheet. A static electric field perpendicular to the graphene will open the band gap in this kind of bilayer graphene. Also a magnetic field modifies the transport on graphene.

6.7.5.2 Fabrication of graphene

Relatively large graphene sheets have been fabricated (up to 100 μm) by mechanical exfoliation of three-dimensional graphite crystals. This method is not easy to control, which makes it too complicated to use for mass production. Once a way of finding the graphene sheet is known, this method will result in high-quality graphene tested by the quantum Hall effect. This material can be used to study the electrical properties of graphene. An improved version of mechanical cleavage is the stamping method in which graphene flakes are transferred using silicon pillars. These methods can be used to fabricate suspended graphene sheets.

Epitaxial growth uses the atomic structure of a substrate to seed the growth of the graphene. Typically it yields a sample with several graphene layers. The interaction between the bottom graphene layer and the substrate has an effect on the properties of the graphene as discussed before.

Another fabrication method is to heat silicon carbide to high temperatures (more than 1100 °C) so that it decomposes to give graphene. Using this method it is difficult to isolate the graphene sheet from the SiC, and it is unlikely to be compatible with fabrication techniques for most electronic applications. Graphene layers have also been fabricated on the surface of ruthenium, Ru.

Graphene can be fabricated by placing graphene oxide paper in a solution of pure hydrazine (a chemical compound of nitrogen and hydrogen), which removes the oxygen functionalities and reduces the graphite oxide paper to single-layer graphene [38]. A safer method is to use hydrazine with water or ethanol.

High-quality sheets of graphene exceeding 1 cm^2 in area have been synthesized via chemical vapor deposition (CVD) on thin nickel layers [39]. These sheets have been successfully transferred to various substrates, demonstrating the method's viability for numerous electronic applications.

6.7.5.3 Graphene transistor

The unique properties of graphene make it a potential candidate to replace silicon and enable high-frequency electronic applications, e.g., fast wireless communication. It will not only replace the connectors of a nanoscale device, but it can be used to replace other

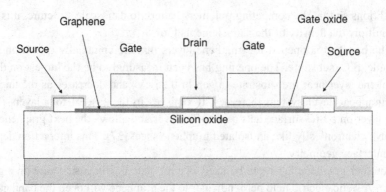

Figure 6.14 Cross-section of a graphene transistor with two channels.

elements of the electronic circuit, e.g., transistors. Fast switching is required in wireless communication; the switching has to occur within the time it takes the signal to pass from one side of the switch to the other. This is usually done by reducing the size of the transistor or making the electron move more quickly. Making conventional transistors smaller than 40 nm is very difficult on silicon. However, with graphene the physical constraints are not as severe as the electrons are 100 times faster than in silicon.

Lemme *et al.* made the first top-gated graphene field effect transistor (GRAFET) [40]. They studied its operation with and without the top gate and found that the top gate decreased the carrier mobility.

The main parts of the GRAFET, shown in Figure 6.14, are: a graphene sheet which acts as a channel, electrodes patterned lithographically to define source and drain contacts, and the local top gate of high *k*-dielectric grown using atom layer deposition (ALD), which makes it possible to build thin layers of the desired material, e.g., aluminum oxide or hafnium oxide. Activation of graphene using NO_2 can be used to prepare the way for the ALD process in order to build the required thin uniform dielectric layer on the graphene surface.

The zero band gap of graphene limits the achievable on–off current ratios. The band gap ($\sim$400 meV) can be widened by cutting graphene into narrow ribbons or using an external electric field as discussed previously. The unique band structure of graphene allows reconfigurable electric-field control of carrier type and density, making graphene a candidate for bipolar nanoelectronics [32]. It is possible to build a top-gated graphene system, a single layer graphene p–n junction in which carrier type and density in two adjacent regions are locally controlled by electrostatic gating. This corresponds to the conventional doping of semiconductors.

In a GRAFET the energy level in the channel is determined not only by the electrostatic capacitance of the gate dielectric but also by the quantum capacitance of graphene, which depends on the density of states, like in the case of nanowires discussed previously.

The cutoff frequency of a top-gated graphene transistor with a gate length of 150 nm has been proven to be 26 GHz [41]. This is ten times slower than the fastest silicon transistor, but terahertz frequency is possible with a gate length of 50 nm. Unlike silicon, graphene is stable and conducts electricity when cut into strips only a few nanometers wide. The carbon–carbon bond is the strongest in nature. The graphene carbon bond holds the structure together even at room temperature where silicon oxidizes and decomposes.

The GRAFET acts like a conventional FET: the measured intrinsic current gain decreases with frequency, f, according to the $1/f$ dependence. This is important when designing circuits based on graphene devices. Real devices require optimization of the materials and the geometry of the system in order to maximize the speed and reduce the leakage.

It is possible to produce a device with a large number of transistors using epitaxial graphene on SiC. De Heer *et al.* have fabricated an array of transistors using standard microelectronics methods [42].

Cheianov *et al.* have claimed that interfaces between materials can be used to manipulate electron flow – just as for light in optics [43]. They predicted the focusing of electric current by a single p–n junction in graphene. The n-type region of graphene can be built by introducing a positive gate-voltage, and the p-type with a negative gate voltage. The ability to focus electric current with p–n junctions on graphene allows the engineering of eletronic lenses. They have shown that precise focusing can be achieved by fine tuning the densities of carriers on the n- and p-sides of the junction to equal values, while the current distribution in junctions with different densities resembles caustics in optics.

Traversi *et al.* have reported the first working device based on graphene: an integrated complementary graphene inverter [44]. Their digital logic inverter uses one p- and one n-type graphene transistor integrated on the same sheet of monolayer graphene.

6.7.5.4 Graphene-based NEMS

As in the case of CNTs, graphene has low mass, high stiffness, and an almost defect-free structure. Small dimensions correspond to high resonance frequencies. With CNT resonators one problem is with the contacting, which leads to a small quality factor, Q. With graphene, the contacts are also challenging, but the problem seems to be smaller than with CNTs. The specific properties of graphene – a strong two-dimensional system with ballistic conductivity, the ability to tune the conductivity externally, the possibility of a defect-free structure and low mass – encourage us to study how good an electromechanical resonator we can build using graphene, and whether we can find novel solutions for the operational principle of the resonator based on these properties.

Bunch *et al.* have built an electromechanical resonator from a graphene sheet suspended over a trench in silicon oxide. The single- and few-layer graphene sheets were manufactured by the exfoliation method. The resulting suspended graphene sheets were described as micrometer-scale cantilevers clamped to the surface by van der Waals forces. The resonator was actuated by applying a voltage between the suspended graphene sheet and the substrate, creating an electrostatic force. The resonance frequencies of the device varied from 42 to 70.5 MHz, for resonators of about 1 mm length. According to the theory of elastic plates the fundamental resonance frequency is proportional to Young's modulus, the thickness of the cantilever, and tension. It is inversely proportional to the mass density and the length of the cantilever. For low tension it scales to h/l^2, where h and l are the thickness and the length, respectively of the cantilever. Thick resonators with tension lead to high resonance frequency. Tension could be used to tune the frequency, but controlling it is challenging. The quality factors obtained at room temperature (from 50 to 800) were not impressive when compared to the Q values of similar volume diamond NEMS devices (2500–3000). Quality factor may be increased by coupling resonators, having several – from one to ten – layers of graphene, and optimizing the geometry and contacting of the resonator. Quality factors in graphene are on the order of 2000 at 10 K and 100 at room temperature [45]. In silicon nitride beam resonators under strong tensile stress, it has been shown that Q factors of 100 000 can be achieved at 10 MHz frequency and as high as 207 000 at room temperature for RF in the case of string resonators [46]. For silicon NEMSs it has also been demonstrated that in doubly clamped beams the Q factor displays $1/f$ frequency dependence, but in the case of graphene there is no clear dependence of Q on thickness or frequency.

In the work of Bunch *et al.*, optical detection was employed to record the oscillation of graphene resonators. Optical detection of graphene resonators becomes increasingly problematic with frequencies above 100 MHz. It is not yet clear what is the best method for read-out gigahertz resonators, but purely electrical measurements, made in the spirit of microwave quantum capacitance measurements [47], are very promising. In such measurements, it is crucial to have moderately large capacitance to ground/gate ($\sim$200 aF), which is easier to obtain in graphene resonators than in their nanotube counterparts.

Unlike Si resonators, the graphene resonator shows no degradation of Young's modulus with decreasing thickness. Graphene – unlike diamond and most other materials – stays strong and stiff even as a single layer of atoms. These mechanical properties make graphene suitable for ultrasensitive force and mass sensing applications. Mass sensing using a cantilever system relies on detecting shifts in the frequency of vibration in response to mass. By using a single layer of graphene atoms as the cantilever, the ultimate limit of two-dimensional nanoelectromechanical systems, a sensor could detect ultrasmall masses. Graphene is also robust enough to resist long-term wear, improving the lifetime and cost efficiency of such balances.

Research questions that have to be considered when proposing graphene resonators for RF applications include: frequency repeatability, fabrication of graphene resonator systems, mass manufacturing, high impedance levels and matching, and thermal stability. In order to predict the performance of the graphene-based electromechanical resonator we have to understand both the mechanical and transport properties of graphene and the coupling of the phenomena. Design methods for predicting the performance of the nanoscale graphene structure are following ab initio and tight binding methods to calculate the energy levels and transport properties of the graphene sheet. Semiempirical models – just like in the case of CNTs – are used to estimate the properties of the material. The estimated values of the material properties are used in macroscopic models based on finite element methods.

6.7.6   Model for design of the nanocomponents

Practical devices are usually complex systems with more than ten parameters that require numerical simulations to be understood properly. In order to predict the performance of a mobile phone we have to combine electromagnetic, acoustical, mechanical, and thermal simulations. This process includes computer aided design (CAD) modelling of the geometry, boundary conditions, and the laws of physics that have to be obeyed. Energy losses and nonlinear phenomena bring more challenges for the prediction of the performance of the systems. The lack of computer power forces us to make approximations and to simplify our system to be able to predict the phenomena. A device with functional nanoscale elements is even more challenging. It requires multiscale modeling of combined macroscale, mesoscale, and nanoscale phenomena.

The decreased size of the nanoscale element and the corresponding change of the effective theory means that scaling the continuum physics is not enough, but further studies are needed to know the limits of current methods and to add quantum mechanics and molecule dynamics to the modeling tool kit in order to design nanodevices. One single model of a device is still a long way off: different descriptions for different regions of the device are needed, e.g., for the contacts, combined with semiempirical methods for the device itself.

In many cases the methods have to be customized in order to find a middle way between the full multi-scale model and the conventional continuum model that describes the macroscopic system, e.g., the mechanical system of curved plates, thin shells, beams, or vibrating rods. Physical quantities related to continuum physics are not always relevant in nanoscale. For example, normally continuum-based (macroscopic) theories are valid to predict the mechanical properties of large CNTs. However, this is not obvious. The elastic parameters for a single-wall CNT depend on such macroscopic parameters as the cross-sectional surface area and the effective thickness of the wall, which are not well defined but which have experimentally defined values, e.g., the effective thickness of a single-wall CNT is defined as 0.34 nm.

The concepts of elasticity and Young's modulus come from continuum, and treating real two-dimensional systems may require something else. Instead of using the universal value for Young's modulus different Young's moduli are required for tension/compression and bending behavior, because tension and compression are governed by the in-plane sigma bond, while pure bending is affected mainly by the out-of-plane pi-bond. More than the classical theory is needed in the case of bent or twisted chiral nanotubes.

One method that researchers have employed is to give an effective thickness for the surface and use the three-dimensional model. Arroyo and Belytschko implemented a finite element method for the nonlinear mechanics of crystalline sheets and nanotubes [48]. The theory generalizes standard crystal elasticity to curved monolayer lattices. The constitutive model for a two-dimensional continuum deforming in three dimensions (a surface) is written in terms of the underlying atomistic model. The resulting hyperelastic potential depends on the stretch and the curvature of the surface, as well as on the internal elastic variables describing the rearrangement of the crystal within the unit cell.

6.8 Future directions

In the world of embedded electronics and augmented reality people will be connected to each other and the environment, where there will be memory and computing everywhere. We will have different ways to communicate, not only by radios based on different frequency ranges of the electromagnetic fields, but also using other interactions; the old means of communication, such as voice, will be seen from a new perspective. It will be a challenge for service and device producers to make the increasingly complex communication system transparent for users, to ensure high performance, and to make devices so easy to use that the user can forget the device and its functionalities. What are the limits according to laws of physics? How can we produce energy-efficient, reliable, fast devices based on new technologies, e.g., new functional materials? How do we integrate new kinds of components in the communication device? How do we ensure that the device and its manufacture is safe and environment friendly? What is the optimal interaction between the device and the user?

We have two kinds of approaches to studying how to develop future communication devices, the gateways between the user, the local environment, and the Internet: top down and bottom up. The point of view of the system is important in order to enable architectural conformity and mass production. Also evolution of today's devices and manufacturing methods will take place more easily than the introduction of novel materials that are the result of nanotechnology because of the disruption these would cause. The working principle of the system of nanocomponents will be changed due to the various effective laws of physics. This may have an impact on the architecture and the system design of the mobile device.

In parallel with the system design we need to develop methods to fabricate, tailor, and image novel materials in order to continue enhancing the properties of the components and the devices. Graphene is one example of a "new" material that might enable both an enhanced and a new kind of communication between people and the environment. The strong carbon–carbon bond makes it the strongest known material. The symmetry of the graphene lattice leads to its exceptional electronics states, which give new possibilities to control the system. Equally any other nanoscale material is of interest when developing new kinds of devices. The novel properties have their basis in theory and new phenomena can be formulated with equations. Current devices are often based on the available materials and on simplified models of the system. By modeling a more complicated system and taking into account the multi-physics phenomena on macro-, meso-, and nano-scale, we will get more design space.

Acknowledgments

The authors would like to thank Romain Danneau, Ilkka Hakala, Pertti Hakonen, Yvette Hancock, Samiul Haque, Visa Koivunen, Leo Kärkkäinen, Reijo Lehtiniemi, Mikko Paalanen, Pirjo Pasanen, Jukka Reunamäki, and Mikko Uusitalo for useful discussions, valuable comments, and suggestions.

References

[1] C. E. Shannon, Communication in the presence of noise, *Proc. IRE*, **37**, 10–21, Jan. 1949.
[2] J. Mitola and G. Q. Maguire, Cognitive radio: Making software radios more personal, *IEEE Pers. Commun.*, **6**, 13–18, 1999.

[3] S. Haykin, Cognitive radio: Brain-empowered wireless communications, *IEEE J. Sel. Areas Commun.*, **23**, 201–220, 2005.

[4] B. Razavi, *RF Microelectronics*, Prentice Hall, 1998.

[5] H. Zimmermann, OSI reference model the ISO model of architecture for open systems interconnection, *IEEE Trans. Commun.*, **28**, 425–432, 1980.

[6] R. H. Walden, Analog-to-digital converter survey and analysis, *IEEE J. Sel. Areas Commun.*, **17**, 539–550, 1999.

[7] P. Eloranta et al., Direct-digital RF-modulator: A multi-function architecture for a system-independent radio transmitter, *IEEE Commun. Mag.*, **46**, 14–151, 2008.

[8] K. van Berkel et al., Vector processing as an enabler for software-defined radio in handheld devices, *EURASIP J. Appl. Signal Processing*, issue 16, 2613–2625, 2005.

[9] A. A. Abidi, The path to the software-defined radio receiver, *IEEE J. Solid-State Circuits*, **42**, 954–966, 2007.

[10] U. Ramacher, Software-defined radio prospects for multistandard mobile phones, *IEEE Computer*, **40**, 62–69, 2007.

[11] C. Kocabas et al., Radio frequency analog electronics based on carbon nanotube transistors, *Proc. Nat. Acad. Sci.*, **105**, no. 5, 1405–1409, 2008.

[12] S. Datta, *Quantum Transport*, Cambridge University Press, 2005.

[13] P. Burke, AC performance of nanoelectronics: towards a ballistic THz nanotube transistor, *Solid-State Electronics*, **48**, 1981–1986, 2004.

[14] P. Burke et al., Quantitive theory of nanowire and nanotube antenna performance, *IEEE Trans. Nanotechnology*, **5**, no. 4, 314–334, 2006.

[15] S. Hasan et al., High-frequency performance projections for ballistic carbon-nanotube transistors, *IEEE Trans. Nanotechnology*, **5**, no. 1, 14–22, 2006.

[16] D. Wang, Ultrahigh frequency carbon nanotube transistor based on a single nanotube, *IEEE Trans. Nanotechnology*, **6**, no. 4, 400–403, 2007.

[17] M. Reznikov et al., Temporal correlation of electrons: suppresson of shot noise in a ballistic quantum point contact, *Phys. Rev. Lett.*, **75**, no. 18, 3340–3343, 1995.

[18] C. Caves, Quantum limits on noise in linear amplifiers, *Phys. Rev. D*, **26**, no. 8, 1817–1839, 1982.

[19] K. Jensen et al., Nanotube radio, *Nano Lett.*, **7**, 374, 2007.

[20] M. P. Anantram and F. Leonard, Physics of carbon nanotube electronic devices, *Rep. Prog. Phys.*, **69**, 507–561, 2006.

[21] J. F. Davis, *High-Q Mechanical Resonator Arrays Based on Carbon Nanotubes*, IEEE, 2003.

[22] J. Kinaret et al., A carbon-nanotube-based nanorelay, *Appl. Phys. Lett.*, **82**, no. 8, 1287–1289, 2003.

[23] V. Sazonova et al., A tunable carbon nanotube electromechanical oscillator, *Nature*, **431**, 284–287, 2004.

[24] H. Peng et al., Ultrahigh frequency nanotube resonators, *Phys. Rev. Lett.*, **97**, 087203, 2006.

[25] P. Ikonen, Artificial electromagnetic composite structures in selected microwave applications, Dissertation, Helsinki University of Technology, 2007.

[26] P. R. Wallace, The band theory of graphite, *Phys. Rev.*, **71**, 622–634, 1947.

[27] K. S. Novoselov et al., Electric field effect in atomically thin carbon films, *Science*, **306**, 666–669, 2004.

[28] Y. Zhang et al., Experimental observation of the quantum Hall effect and Berry's phase in graphene, *Nature*, **438**, 201, 2005.

[29] S. V. Morozov et al., Strong suppression of weak localization in graphene, *Phys. Rev. Lett.*, **97**, 016801, 2006.

[30] J. S. Bunch, *et al.*, Impermeable atmoc membranes from graphene sheets, *Nano Lett.*, 2008.

[31] M. Y. Han *et al.*, Energy band-gap engineering of graphene nanoribbons, *Phys. Rev. Lett.*, **98**, 206805, 2007.

[32] J. R. Williams, L. DiCarlo, and C. M. Marcus, Quantum Hall effect in gate-controlled p–n junction of graphene, *Science*, **317**, 638–641, 2007.

[33] S. Y. Zhou *et al.*, Substrate-induced bandgap opening in epitaxial graphene, *Nature Mater.*, **6**, 770, 2007.

[34] H. Min *et al.*, Intrinsic and Rashba spin-orbit interactions in graphene sheets, *Phys. Rev. B*, **74**, 165310, 2006.

[35] C. H. Park *et al.*, Anisotropic behaviours of massless Dirac fermions in graphene under periodic potential, *Nature Phys.*, **4**, 213–217, 2008.

[36] R. Ruiz *et al.*, Density multiplication and improved lithography by directed block copolymer assembly, *Science*, **321**, 936–939, 2008.

[37] F. Varchon *et al.*, Electronic structure of epitaxial graphene layers on SiC: effect of the substrate, *Phys. Rev. Lett.*, **99**, 126805, 2007.

[38] V. C. Tung *et al.*, High-throughput solution processing of large-scale graphene, *Nature Nanotech.*, **4**, 25–29, 2009.

[39] K. S. Kim *et al.*, Large-scale pattern growth of graphene films for stretchable transparent electrodes, *Nature*, **457**, 706–710, 2009.

[40] M. C. Lemme *et al.*, A graphene field effect device, *IEEE Electron Dev. Lett.*, **28**, 282–284, 2007.

[41] Y.-M. Lin *et al.*, Operation of graphene transistors at GHz frequencies, *Phys. Rev. B*, 2008.

[42] J. Kedzierski *et al.*, Epitaxial graphene Transistors on SiC substrates, *IEEE Trans Electron Dev.*, **55**, 2078–2085, 2008.

[43] V. V. Cheianov, V. Fal'ko, and B. L. Altshuler, The focusing of electron flow and a veselago lens in graphene pn-junctions, *Science*, **315**, 1252, 2007.

[44] F. Traversi, V. Russo, and R. Sordan, Integrated complementary graphene inverter, arXiv:0904.2745v1 [cond-mat.mes-hall] 17 Apr 2009.

[45] J. S. Bunch *et al.*, Electromechanical resonators from graphene sheets, *Science*, 26, 490–493, 2007.

[46] S. S. Verbridge *et al.*, High quality factor resonance at room temperature with nanostrings under high tensile stress, *J. Appl. Phys.*, **99**, 2006.

[47] M. A. Sillanpää *et al.*, Direct observation of Josephson capacitance, *Phys. Rev. Lett.*, **95**, 206–806, 2005.

[48] M. Arroyo, and T. Belytschko, Finite element method for the non-linear mechanics of crystalline sheets and nanotubes, *Int. J. Num. Meth. Eng.*, **59**, 419–456, 2004.

7 Flat panel displays

A. J. Flewitt and W. I. Milne

7.1 Introduction

A distinguishing feature of mobile telephones compared with "fixed line" devices since their widespread take-up has been in the inclusion of a display. Given that the functionality of these early devices was largely limited to making and receiving voice calls, the need for a display is not immediately apparent, and indeed many "fixed line" telephones still do not include even the most basic of displays. However, it is the inclusion of a high-quality display that has enabled the mobile telephone to be transformed into a more general mobile communication device. This is because voice communication is an inherently slow means of receiving information (although it is much faster than writing or typing in terms of transmitting information). Human beings absorb information that is written down much more quickly and a wider diversity of information can be presented, such as menu icons, maps, and photographs. In many senses, the nature of the display that can be incorporated into a mobile device limits the functionality that can be attained, and much of the desired functionality of future mobile devices that move us towards a regime of ubiquitous electronics will require significant development of display technology.

This chapter will examine the history of display technology with reference to mobile devices, consider technology developments that are designed to move displays towards greater robustness, improved resolution, higher brightness, enhanced contrast, and mechanical flexibility, and look forward to the "grand problems" that future display manufacturers will need to consider. As we will see, nanotechnology plays a ubiquitous role in enabling both current and future displays, and at the end of the chapter we shall review this role explicitly.

7.2 The emergence of flat panel displays

7.2.1 The dominance of the cathode ray tube (CRT)

The first 60 years of display technology was essentially dominated by the CRT. In the CRT, a beam of electrons is created by extracting electrons from a heated cathode by

Nanotechnologies for Future Mobile Devices, eds. T. Ryhänen, M. A. Uusitalo, O. Ikkala, and A. Kärkkäinen. Published by Cambridge University Press. © Cambridge University Press 2010.

thermionic emission, and accelerating these electrons by applying a high voltage to a set of electrodes. A further set of electrodes is used to scan the electron beam over a phosphor-coated screen. The phosphors then emit light due to the impact of the electrons. This entire process takes place under a vacuum. An input video signal is decoded, and used to modulate the intensity of a beam of electrons so that an image is produced on the screen. Decades of research into this technology means that the resulting image is of an exceptionally high quality in terms of resolution, brightness, contrast, viewing angle, and speed. Consequently the television market is still led by CRT technology, and the market share for CRT televisions is only expected to drop below 50% for the first time in 2009. However, the requirement for a high voltage to accelerate and guide the electrons together with the need for a long vacuum tube through which the electrons are scanned means that this technology is unsuited to mobile display applications (although even here, some very small-scale CRT displays have been developed for very specialized applications where cost is not a key issue, such as helmets with integrated displays for military use). Furthermore, the CRT requires the use of significant quantities of heavy metals which are environmentally unfriendly.

CRTs show very clearly the three key elements that are required for a full display: a source of light, a means of modulating the light intensity with time over a display surface, and a method of controlling this modulation from a video signal. Any new display technology for mobile applications must be able to integrate all three of these elements in a lightweight, flat panel package with low power consumption and without the need for high-voltages whilst maintaining a high-quality picture.

This has proved to be very challenging. However, a clear leading technology has emerged in the form of liquid crystal displays (LCDs), which currently dominate the mobile display market. The rate of improvement in LCDs since the 1990s has been very impressive, and in the next section, we will look at how LCDs work, but we will also see that they too have limitations; this has seen a challenger emerge in the form of displays based upon organic light emitting materials, and we will also consider this approach along with other new display technologies based upon field emission, electrophoretic systems, and electroluminescent nanoparticles.

7.2.2   The rise of LCDs

LCDs have dominated the mobile flat panel display market since their commercial emergence in the 1970s. In essence, liquid crystals are long-chain molecules – typically ~ 3 nm long and ~ 1 nm in diameter – which find it energetically favorable at room temperature to align themselves in an ordered structure. In this regard, they are true nanomaterials, although their widespread use came about long before this term emerged, and so they are frequently overlooked when considering the use of nanomaterials today. The resulting material exhibits optical anisotropy. This is utilized by forming a thin film of the liquid crystal between two transparent electrodes (normally made from indium tin oxide) as shown in Figure 7.1. The surface of the electrodes can be prepared such that the layer of liquid crystal in contact with the surface lines up in a particular direction. In the simplest form of twisted nematic device, ambient light enters the system through a top

(a)   (b)

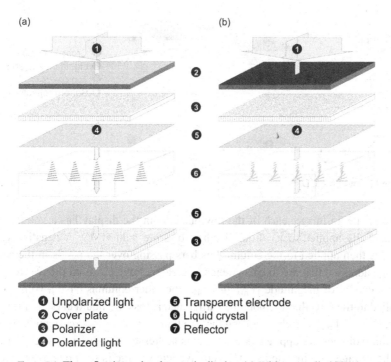

❶ Unpolarized light   **❺** Transparent electrode
❷ Cover plate   **❻** Liquid crystal
❸ Polarizer   **❼** Reflector
❹ Polarized light

Figure 7.1 The reflective twisted nematic display. (a) With no applied bias between the electrodes, the orientation of the liquid crystal is such that polarized light is twisted to allow reflection from the bottom of the cell. (b) With an applied bias between the electrodes, the liquid crystal molecules align vertically so that no change in light polarization occurs and no light is reflected.

polarizer so that the light entering the liquid crystal is plane polarized. The liquid crystal is oriented such that the top and bottom layers are at 90° to each other. This causes the polarization of the light to be rotated through 90° also before passing through a second polarizer that is at 90° to the first. The light then hits a back reflector before traveling back through the system and out. By applying a voltage across the two electrodes, the liquid crystal can be made to reorient so that the liquid crystals attempt to line up parallel to the electric field. As a result, the liquid crystal no longer acts to rotate the polarization of the light, and so all the light will be blocked by the crossed polarizers and none will be reflected. By creating a simple pattern of electrodes over a surface, a simple display can be created, such as those found in digital watches. However, as contacts need to be made to each independent element of the display, only very crude data can be displayed.

In order to increase the complexity and variety of the data that can be displayed, it is necessary to divide the display surface into an array of square pixels. For even a moderately sized array, it is impossible to make a contact to each individual pixel, and so passive matrix addressing is used. Passive matrix liquid crystal displays (PMLCDs) take advantage of the fact that a minimum (threshold) rms voltage V_{lc} must be applied across the liquid crystal in order to induce the reorientation. The display is constructed such that a series of conductive row bus bars pass over the top of the liquid crystal and a series of column bus bars pass underneath (or vice versa) as shown in Figure 7.2.

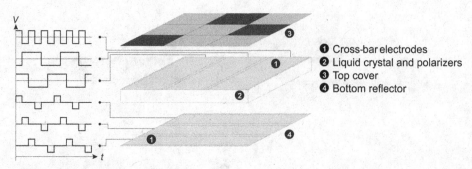

Figure 7.2 The PMLCD.

To create an image, each of the row bus bars in the display has a voltage of $+V_{\text{row}}$ sequentially applied to it while all other rows are held at $0\,\text{V}$. A negative voltage of $-V_{\text{col}}$ is then applied to the column bus bars passing over the pixels in the row that is being addressed that are to be switched to create a large potential difference across the liquid crystal of magnitude $|V_{\text{row}} + V_{\text{col}}|$. The other columns have a voltage of $+V_{\text{col}}$ applied to them so that the potential difference across the liquid crystal is significantly less, $|V_{\text{row}} - V_{\text{col}}|$, thereby ensuring that the liquid crystal is not reoriented. Appropriate column voltages are applied as each row is addressed. On the next sweep of rows, all of the voltages are inverted so that zero average voltage is seen by a pixel over a long period of time.

This approach has the advantage of allowing a very simple display construction and addressing scheme. However, it has a major problem when it comes to having a large number of rows which can be seen by considering the rms voltage seen by pixels on the "on" state, which is

$$\langle V_{\text{on}}^2 \rangle = \left[\frac{(n-1)}{n} V_{\text{col}}^2 + \frac{1}{n} (V_{\text{row}} + V_{\text{col}})^2 \right] \tag{7.1}$$

and that seen by pixels in the "off" state, which is

$$\langle V_{\text{off}}^2 \rangle = \left[\frac{(n-1)}{n} V_{\text{col}}^2 + \frac{1}{n} (V_{\text{row}} - V_{\text{col}})^2 \right] \tag{7.2}$$

where n is the number of rows. The ratio of these two voltages is therefore given by

$$\frac{\langle V_{\text{on}}^2 \rangle}{\langle V_{\text{off}}^2 \rangle} = \frac{(n-1)V_{\text{col}}^2 + (V_{\text{row}} + V_{\text{col}})^2}{(n-1)V_{\text{col}}^2 + (V_{\text{row}} - V_{\text{col}})^2}, \tag{7.3}$$

which tends towards 1 as n becomes large. In other words, the difference between the "on" state and "off" state voltages becomes very small, resulting in decreasing contrast and increased cross-talk between pixels. Therefore, this type of display also has a limited data content.

As a consequence, active matrix liquid crystal displays (AMLCDs) have been developed, a cross-section of one of which is shown in Figure 7.3. A bottom polarizer produces plane polarized light from a diffuse light source that is located behind the display. An array of transparent electrodes is patterned on a lower glass plate through which the

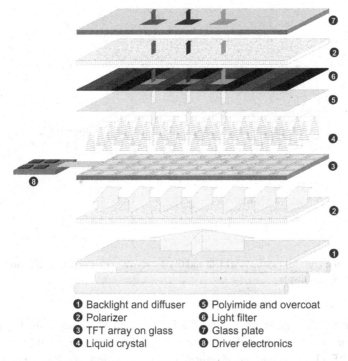

● Backlight and diffuser   ● Polyimide and overcoat
● Polarizer   ● Light filter
● TFT array on glass   ● Glass plate
● Liquid crystal   ● Driver electronics

Figure 7.3 Schematic diagram of the AMLCD.

polarized light passes. A thin film transistor (TFT) is located at each of these pixels. Figure 7.4(a) shows the pixel circuit. As with the PMLCD, each pixel is served by a set of row and column bus bars. The column bus bars carry the voltage required to program each pixel. To program a particular row of pixels, a high voltage is applied to the row bus bar, $V_{address}$, which turns on transistor T1, causing the pixel electrode to see the programming voltage, V_{data}, that has been applied to the column bus bar. This voltage then determines the degree of rotation (and hence light transmission) of the liquid crystal at each pixel which is sandwiched between this TFT array and a single, large-area, transparent electrode which covers the whole display and is held at 0 V. Color is added to the display by a set of color filters.

The key advantage of this scheme is that the liquid crystal acts as a capacitor, and so, as long as the TFT has a very low off-state current, the voltage applied to the liquid crystal will remain constant while other rows of the display are addressed, and will be unaffected by the column voltages used to program other rows of the display. As a result, cross-talk is minimized and contrast ratio maximized. The limit on the size of the display now becomes the resistance of the bus bars, either which must be increased in width, or a more highly conductive material must be employed, such as copper, in order to reduce the resistance.

Producing this layer of TFTs itself presents a major challenge. A 1680 × 1050 pixel display with a red, green, and blue sub pixel at each point requires over 5 million TFTs. This is achieved using thin film silicon technology. In the latest generation of display

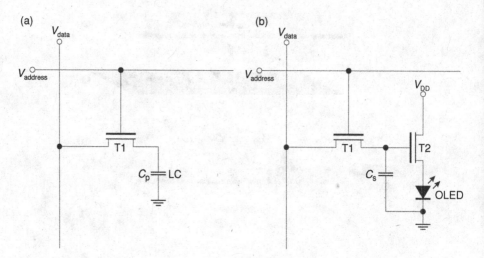

Figure 7.4 Pixel circuit for (a) the AMLCD and (b) the active matrix organic light-emitting display (AMOLED) display.

manufacturing lines, glass substrates of 2.16×2.4 m dimensions are loaded into a vacuum chamber and are heated to a temperature between 200 and 350 °C. In a process known as RF plasma-enhanced chemical vapor deposition (RF-PECVD), silane gas (SiH_4) is then injected into the chamber and excited by application of a 13.56 MHz oscillating electric field. The activated gas then produces a highly uniform thin film of semiconducting hydrogenated amorphous silicon (a-Si:H) on the surface of the substrate. Doping can be achieved by adding phosphine (PH_3) which results in phosphorous donors being incorporated into the a-Si:H [1]. Insulating silicon nitride can be produced by adding ammonia (NH_3) diluted in hydrogen or nitrogen to the silane. Photolithography can then be used to pattern the materials to form the TFTs and the bus bars.

While a-Si:H technology has been highly successful, it does present limitations due to the fact that the amorphous structure of the a-Si:H means that its electronic properties are significantly inferior to those of crystalline silicon. For example, the field effect mobility, which influences both the maximum current that can be driven as well as the switching speed of the transistors, is $\sim$1000 times lower compared to crystalline silicon. Consequently, while the a-Si:H can be used for transistor T1 in the pixel circuit of Figure 7.4(a) where the switching speed is slow, it cannot be used to decode a video signal to drive the row and column bus bars. Crystalline silicon integrated circuits must be used for this, and interconnects employed to connect these to the bus bars, as shown in Figure 7.3. This is a source of mechanical weakness which limits the durability of the display overall, particularly in mobile applications. The electrical properties of the silicon thin film can be enhanced by a factor of $\sim$100 by crystallizing the a-Si:H to form polycrystalline silicon. However, this process requires either high temperatures or the use of an excimer laser beam to induce the phase change, both of which are costly, and so polycrystalline silicon tends only to be used on very high-quality, small displays, such as for handheld video cameras. Furthermore, the need to heat the substrate during a-Si:H deposition means that glass must be used rather than plastic, so the display is inflexible and prone to shatter on impact.

AMLCDs have a further key limitation for mobile applications in that most of the power consumed by such a display goes into the backlight. Four technologies are commonly used for AMLCD backlights: halogen filament lamps, electroluminescent foils, light emitting diodes, and cold cathode fluorescent lamps. Of these, the last is the most energy efficient, typically consuming $\sim$50 W per square meter of emitting surface with a luminescence of 100 cd/m^2. However, the majority of this light never actually leaves the display, as the principle of operation is to remove light through the action of both the liquid crystal and the color filters. Not only does this lead to high power consumption, but it also reduces the contrast ratio, as the black areas of a picture are not really black, with some light emission. Therefore, these displays are more taxing on the eyes and more difficult to see when there is significant background illumination, as is usually the case outdoors during the day.

Given the importance of low power consumption to maximize battery life or minimize battery size in mobile devices together with the desire for more durable and flexible displays based on plastic substrates with better picture quality, alternative technologies have been proposed to replace both the a-Si:H backplane electronics and the liquid crystal. In the next section, we shall look at three new flat panel display technologies that are currently being developed. However, as we shall see, displacing an incumbent, dominant technology requires the new technology to be both significantly better and also cheaper to manufacture.

7.3 New technologies for flat panel displays

7.3.1 Field emission displays (FEDs)

FEDs are not new, having first been muted about 50 years ago, but only since in the mid–late 1990s have they received significant attention as a potential alternative to AMLCD technology. The aim was to address three major problems with AMLCDs of the time: their high power consumption and limited contrast (both of which continue to be an issue), and their limited viewing angle (which has largely been resolved through the use of in-plane switching of the liquid crystals). It was proposed that this could be achieved by returning to phosphor technology, as found in the CRT, for light emission.

The structure of the FED consists of a front glass panel that is coated in phosphor pixels which emit red, green, or blue light upon impact by high-energy electrons. However, rather than having the single thermionic emitter of electrons and the bulky scanning tube of the CRT, electrons are emitted locally from a flat surface a short distance away from the phosphor screen by field emission, as shown in Figure 7.5.

The field emission process from surfaces was studied in detail by Fowler and Nordheim [2]. In a metal, a barrier exists that prevents electrons from leaving the material and entering free space which is defined by the work function. In an n-type semiconductor, this barrier becomes the electron affinity. When a high electric field is applied to the

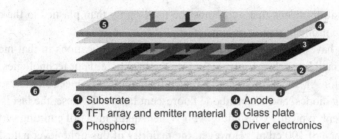

❶ Substrate   **❹** Anode
❷ TFT array and emitter material   **❺** Glass plate
❸ Phosphors   **❻** Driver electronics

Figure 7.5 Schematic diagram of the FED.

surface of the material, this barrier is reduced in both width and height such that it is possible for electrons to quantum mechanically tunnel into free space with a current density, J, given by

$$J = \frac{\alpha A (\beta F_a)}{\phi} \exp\left(\frac{-b\phi^{3/2}}{\beta F_a}\right),$$

$$(7.4)$$

where ϕ is the barrier height with no applied field, F_a is the applied field, A is the ratio of the electron emission area relative to the anode area, and β is a geometrical field enhancement factor (equal to 1 for a flat surface). a and b are given by fundamental constants,

$$a \equiv \frac{e^3}{16\pi h}; \; b \equiv \frac{4\sqrt{2m_e}}{3e\hbar},$$

$$(7.5)$$

where e is the electronic charge, m_e is the electronic mass, and $\hbar$ is the Plank constant divided by 2π.

For mobile applications, it is not acceptable to have high voltages present, and this has two major consequences for the FED. Firstly, the phosphor must emit light even though the bombarding electrons have very low energies, and so a range of low-voltage phosphors were developed to meet this need which could operate for biases of a few tens of volts. Second, it must be possible for significant field emission to occur for very low applied electric fields (~1 V/μm). One approach to maximizing electron emission for low applied fields is to use emitting materials that have a low barrier, ϕ. Carbon thin-film materials were considered attractive for this purpose as diamond has a very low electron affinity, which even becomes negative if terminated with hydrogen [3]. However, a second, and more effective approach, is to use the phenomenon that an electric field can be locally enhanced by a conducting point, and it is for this reason that the geometric field enhancement factor, β, must be included in (7.4). Such geometrical enhancement can be achieved by introducing electronic inhomogeneity into a homogeneous thin film, such as by the inclusion of conducting sp^2-rich regions in an otherwise high sp^3-content diamond-like carbon film [4]. Alternatively, nanotechnology can be used to create a geometrically inhomogeneous material, and carbon nanotubes are ideal for this purpose, having an exceptionally high aspect ratio and high conductivity. An optimum density of carbon nanotubes exists. Clearly, the higher the surface density of carbon nanotubes, the greater the number of field emission sites. However, if the number density is too high, then a screening effect takes place which acts to reduce β [5].

Although FEDs have received significant industrial backing, and good quality demonstrators have been produced, most notably by Samsung, no displays based on this technology are currently on the market. An inherent problem with FEDs is the need for a vacuum between the anode and the cathode. Degradation of the vacuum results in ionization of the residual gas by the emitted electrons and poisoning of the emitting material resulting in degraded performance. The difficulty in obtaining a robust, well-packaged display with a long lifetime may be one reason why the FED has not succeeded commercially. However, the main problem is almost certainly the success of the AMLCD and that improvements in display quality have reduced the potential market for FEDs to such an extent as to make them commercially unattractive. One application for field emission technology may, however, be to act as a large-area, uniform backlight for AMLCDs. In this case, there are less stringent requirements for patterning of both the emitter and the phosphor and greater emitter redundancy can be built into the display. The result may be a backlight that has a lower power consumption than the cold cathode fluorescent lamp, but even here there is severe competition from the development of light emitting diode backlights.

7.3.2 Organic light emitting displays (OLEDs)

Organic electroluminescent materials have also received significant attention as a potential flat panel display technology since the late 1980s. The motivation is the potentially simpler structure of such a display compared to AMLCD. OLEDs work by having a patterned array of red, green, and blue electroluminescent material covered in a transparent conductor which acts as a common anode. On the other side of the organic material is an array of patterned cathode electrodes to which a voltage can be applied by an active matrix (AM) backplane. By injecting holes into the organic material at the anode and electrons at the cathode, exitons are created which emit light upon recombination. To aid this process, additional layers are also usually incorporated to aid carrier injection and transport.

In addition to the simplicity of the structure of the display, AMOLEDs have three other significant advantages over AMLCDs. Firstly, the light emission process is highly efficient, and all the light that is emitted is used to create the display image (unlike in AMLCDs where much of the light from the backlight is filtered out). Consequently, the display consumes very little power, making it ideal for mobile device applications. Secondly, because there is no liquid crystal, the viewing angle is very high. Finally, the use of polymers potentially allows both the production cost to be reduced and a flexible display to be manufactured, as polymers are far less prone to cracking under stress than inorganic materials.

Two classes of organic electroluminescent materials have emerged: small-molecule materials [6, 7] and polymers [8, 9]. The advantage of the polymers is that they can be solution processed, while small-molecule materials generally require vacuum deposition by evaporation (although some solution-processable small-molecule materials are starting to appear). Small-molecule emitters, on the other hand, tend to have longer lifetimes and improved efficiencies. The greater efficiency is due to the fact that excitons can exist in one of two states: the singlet state, which ~25% of excitons occupy, and the triplet state which is occupied by the remainder. In most organic materials, transitions between these states are very unlikely at room temperature, and so the relative population of singlet and triplet states is effectively "frozen." Quantum mechanical spin conservation forbids radiative recombination from the triplet states, and this places a severe limitation on the efficiency of the polymer emitters. In the small-molecule materials, the presence of a heavy metal ion enables radiative emission from the triplet states, and so the efficiency is increased by a factor of 4.

While OLED technology has received significant investment since the early 1990s, the number of actual displays in production is relatively small, being limited to a few small-area displays for mobile telephone applications, PMLCD backlights, and low-information-density displays (such as the charge indicator on Philips electric shavers). This is indicative of two key problems with OLEDs that remain active issues: limited material lifetime due to oxidation and the AM backplane performance.

Material lifetime is an issue that has dogged OLED technology since its inception, and although red light small-molecule light emitting materials are available with operational lifetimes of ~500 000 hours, achieving lifetimes greater than the ~20 000 hours required for most display applications for blue light emitters is proving to be more challenging. As a red, a green, and a blue emitter is required in order to obtain a full color display, the lifetime of the blue emitter limits that of the whole display. Furthermore, it is important that all three colors degrade at the same rate in order for the display not to take on an unwanted color hue as it ages (although some electronic compensation circuitry may be employed to mitigate this effect).

The performance of the AM backplane is also an issue. Figure 7.4 compares the pixel circuit for the AMLCD with that for the AMOLED. Whereas transistor T1 in the AMLCD has a very low duty cycle and low on-state current requirement, transistor T2 in the AMOLED has to drive large currents through the light emitting material for as long as the pixel is illuminated. TFTs using a-Si:H as the channel material suffer from significant degradation under such conditions, which is manifested as a change in the threshold voltage of the device as a function of time. Two approaches are being adopted to address this issue. One is to increase the complexity of the pixel circuit to introduce a means of compensating for the change in the threshold voltage. This has the advantage of allowing display manufacturers to continue to use well-known a-Si:H technology. However, the complexity of the pixel circuit is significantly increased, which has a significant impact upon yield and cost. The alternative approach is to use a different material for the TFTs. Poly-Si does not suffer from the instability of a-Si:H, but has a high manufacturing cost (as discussed in Section 7.2). Microcrystalline silicon and nanocrystalline silicon are materials that can be produced using the same equipment as a-Si:H. They consist of small crystalline regions of silicon embedded within an amorphous matrix. TFTs fabricated using this material have a significantly reduced threshold voltage shift, but tend to suffer from high off-state leakage currents due to the microstructure of the material. Beyond silicon, amorphous metal oxides, such as indium gallium zinc oxide (IGZO) are being proposed as potential alternatives to thin-film silicon technology, but there is currently insufficient evidence to demonstrate whether the stability of these materials will be sufficient to drive an AMOLED for the required lifetime of the display. Finally, organic semiconductors, such as triisopropyl silyl (TIPS)-pentacene or polydialkyl quater thiophene (PQT), are also receiving attention due to the potential for solution processing of transistors at very low lost. However, these materials also suffer from relatively low carrier mobility and degradation mechanisms of their own. As such, to date no one backplane technology has yet emerged that can deliver an OLED AM backplane with a sufficiently low-cost structure and high performance to enable production.

7.3.3   Electrophoretic displays

Electrophoretic displays using "electronic ink" (e-ink) are a rather different class of display technology that emerged during the late 1990s, and it now dominates as the preferred technology for e-book readers. Developed by E-Ink Corporation in partnership with a number of display manufacturers, e-ink displays are reflective rather than emissive. Rather like the OLED, the construction of the display consists of an AM backplane where each pixel has a large conducting pad to which a positive or negative voltage can be applied. Lying above the backplane is a layer of screen-printed, small, transparent capsules which contain a clear fluid, positively charged white particles, and negatively charged black particles. Once again, we see the clear use of an engineered nanomaterial in a display application. Above the layer of capsules is a common, transparent electrode on a transparent substrate (such as glass or plastic). Applying a positive voltage to a particular pixel relative to the common top electrode attracts the black particles in the capsules above the pixel to the bottom of the capsule, while the white particles will drift to the top. Similarly, applying a negative voltage to the pixel will cause the white particles to drift to the bottom of the capsules and the black particles to the top.

The result is a display that looks very much like a printed page. Consequently, it is very easy on the eye, and can be read for long periods without the eye strain that most emissive displays tend to produce. The display also has a very low power consumption as once the image is formed; the white and black particles will remain at the top and bottom surfaces indefinitely, even when the driving voltage is removed. Furthermore, the capsules are mechanically stable, and so the technology lends itself to flexible and robust displays. It is for these reasons that e-ink has found an immediate niche in e-book readers, and is likely to expand to fill other display applications, particularly where low power consumption is key and an emissive display is not required.

As with OLEDs, the AM backplane technology to drive these displays is being contested by a number of solutions. As a high switching speed is not required, the duty cycle is low, and the on-state current requirement is also low, the main contenders are a-Si:H, amorphous metal oxides and polymer semiconductors. In this case, a-Si:H has the advantage of being a tried-and-tested technology, and PVI are developing a process that uses standard AMLCD production facilities to produce an e-book a-Si:H TFT backplane on a robust, polyimide substrate. However, the low device performance requirements also suit the polymer semiconductors well, and Plastic Logic are also developing an e-book reader at the time of writing using a polymer semiconductor TFT backplane.

7.3.4   Quantum dot light emitting diode (QDLED) displays

Of all the emerging display technologies considered in this section, the quantum dot light emitting diode (QDLED) displays are by far the most nascent. As a technology, it represents a direct challenge to OLEDs. The electroluminescent material consists of a disordered array of nanometer-scale particles with an inorganic core (e.g. CdSe) and shell (e.g. ZnS) [10]. Each of the nanoparticles is surrounded by a layer of organic molecules. The structure of the display is essentially identical to that of the OLED, consisting of a layer of QDLEDs sandwiched between an anode covered with a hole injection layer and a cathode covered with an electron injection layer. Light emission occurs by exciton recombination, either following direct hole and electron injection into the quantum dot or by a resonant energy transfer into the quantum dot from excitons on the organic coating. The wavelength of the light emitted is primarily determined by the dimensions of the quantum dot, and so good control of the uniformity of nanoparticle size is key for ensuring good color purity.

Like organic polymers, QDLEDs can be solution processed, enabling patterning by a variety of printing techniques at low cost. However, it is believed that the color range that can be produced by the QDLEDs will surpass that currently possible with organic electroluminescent emitters, and hence the quality of the display will be superior to OLEDs. QDLEDs also boast high efficiency and so low power consumption and good lifetimes (reports already exist of lifetimes in excess of 50 000 hours, although explicit data for red, green, and blue emitters are not available).

A key to enabling display production using QDLED technology will be the ability to deliver a suitable AM backplane. In this regard, the challenges are similar to those that were discussed in Subsection 7.3.2 for OLEDs as the pixel circuit will be very similar to that for OLEDs shown in Figure 7.4. It is almost certain that a-Si:H TFTs will struggle to deliver the required on-state current and lifetime, and so the other silicon, metal oxide, and polymer semiconductor technologies will compete to provide a low-cost, manufacturable solution.

7.4   Displays as an intuitive human interface

Thus far, we have considered the display of a mobile device in a very traditional sense as the means by which visual information can be conveyed to the user. However, this situation is rapidly changing, and the display is becoming more of an intuitive human interface. This can be seen, e.g., in the Nokia N95 and N97 devices, the Apple iPhone, and the Microsoft Smart Table, where the display is touch-sensitive, enabling the user to manipulate virtual objects in an intuitive fashion, e.g., by simply 'pushing' them over a surface or pulling them to stretch them.

Figure 7.6 Schematic diagram of a capacitive touch sensor system.

The dominant technology for touch-sensitive displays is the capacitive touch sensor. A set of transparent electrodes (usually using indium tin oxide) are patterned on the surface of a display, as shown in Figure 7.6. When a potential difference is applied between two electrodes, an electric field will be set up which, because of the very thin nature of the conducting film, will be dominated by fringing fields that exist outside the plane of the electrodes. The capacitance between the two electrodes is therefore, in part, determined by the dielectric properties of the environment above the display, which is normally air with a relative permittivity $\sim$1. Skin behaves as a dielectric with a significant loss term, and so, when the surface of the display is touched, the capacitance between the electrodes is significantly changed, and the position of contact can be determined.

Such a capacitive touch sensor has the advantage of being very simple to implement on almost any display surface. However, it does not provide any information regarding the force of the contact, and so an extra dimension of information cannot be gathered. Neither does it allow for other objects to touch the display whose dielectric properties are similar to those of air. It is for this reason that alternative technologies are being actively investigated. These include resistance monitoring, infrared light, and surface acoustic wave devices. An alternative solution using the piezoelectric properties of zinc oxide nanowires is being tested through a collaboration between Nokia and the University of Cambridge. In this scheme, an array of vertically aligned zinc oxide nanowires is embedded within a polymer, such as polystyrene, and patterned with an array of top row and bottom column electrodes to produce a passive matrix array, as shown in Figure 7.7. When the surface is touched, the zinc oxide nanowires are deformed within the polymer and the piezoelectric nature of the nanowires means that a charge is produced and voltage developed between the two electrodes. As the deformation of the nanowires (and hence the voltage developed) is dependent on the force applied, it is possible to gain information on the pressure applied during contact, providing additional information to the mobile device.

A further challenge for touch-sensitive displays is that there is currently no haptic (touch) feedback to the user. In many devices, a simple audible "beep" is used to confirm that a touch has been registered. However, a proper keyboard uses the fact that the depression of a key is felt to provide a far more intuitive feedback to the user. Some mobile devices are overcoming this problem by using the existing "silent mode" vibration system to provide a momentary vibration that feels like a "knock." Other solutions, such as that being developed by NXT Technology, use piezoelectric devices mounted around the edge of a display to provide a mechanical "shock wave" to the display surface that again gives a haptic sensation of a "click" to the user.

All current systems apply the haptic signal to the whole of the display and not just to the point of contact, which limits the ability to use the system for "multitouch"

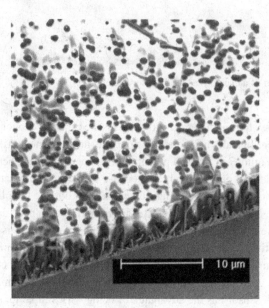

Figure 7.7 A scanning electron micrograph of the vertically aligned zinc oxide nanowires embedded within a polymer.

operations in which contact is made at different points of the display simultaneously. There is significant interest in the ability to provide localized haptic sensation, not only to address the issue of how to provide multitouch haptic feedback, but more generally to allow the display surface to be programmed with "haptic icons" that would allow the user to navigate over the display simply by touch, rather as a touch-typist is able to do with a QWERTY keyboard. Further ahead, it is possible to envisage a display surface that possesses a range of sensors that can detect a variety of signals from the user, including biological information, such as analysis of the user's sweat as a means of assessing wellbeing. Detection of force could even be used to monitor the user's heartbeat at a very simple level. In such a vision, the display is no longer a simple means of conveying visual information to the user – instead it becomes an intuitive interface for the user to access the digital world with a minimal electronic barrier.

7.5   Conclusions

In this chapter, we have charted the development of display technologies for mobile applications from the CRT, through the currently dominant liquid crystal technology, on to emerging technologies such as OLEDs, finishing with the display as a more general human interface with the advent of haptic devices. As with any such short review, there is much that cannot be covered. Not least, we have not considered projection displays. These are currently most common for large-area displays, and are dominated by the digital micromirror technology that has been developed by Texas Instruments. However, other projection technologies are also emerging, such as liquid crystal on silicon (LCOS) in which the Fourier transform of a display image is written in a liquid crystal to induce a localized phase change in a reflected laser beam, which then converts the Fourier transform back into a real space image in the far field. We have also not considered three-dimensional displays. These are most common in cinema viewing and gaming consoles. In order to produce a three-dimensional display, a different image must be projected to each eye. There are two means of achieving this. One is to present two simultaneous images of different colours or light polarizations and for the viewer to wear special glasses that only allow one of the two images to reach each eye. The second is to use special optics to present a different image to each eye directly. Neither is ideal though, either requiring the wearing of special glasses (as in the former case) or having a small viewing angle (as in the latter case). Both of these approaches will therefore be inconvenient for mobile device displays, and much work will be needed to implement a commercial solution.

Although we have not explicitly highlighted the role of nanotechnology in these devices, it should be clear that, in all cases since the CRT, nanotechnology has enabled the development of displays. In the case of liquid crystals, the liquid crystal polymers themselves are undoubtedly nanomaterials, with no dimension exceeding a few nanometers. FEDs will only be enabled by high-aspect-ratio, nanostructured materials that provide a significant field enhancement factor as well as offering high conductivity, such as carbon nanotubes. In many senses, the organic small molecules and polymers used in OLEDs are nanomaterials with at least two dimensions of only a few tens of nanometers. The electrophoretic e-ink uses nanometer-scale white and black particles within the capsules to produce the required reflectivity. Of all the technologies covered, quantum dot light emitters perhaps use the special properties encountered at the nanoscale most explicitly to generate color light emission. Meanwhile, the use of zinc oxide nanoparticles to create a deformable, piezoelectric material for touch sensing shows how nanotechnology can provide new solutions to existing problems.

In conclusion, humans have five senses: touch, sight, sound, smell, and taste. The future of mobile communications will see electronic devices working with these senses as broadly as possible. The challenge for displays is to help to achieve this goal by not only providing a high-quality image, but by allowing as broad an interaction with the user as possible, and it is clear that nanotechnology can, does, and will be essential for achieving this.

Acknowledgments

The authors would like to thank all of the members of the Electronic Devices and Materials Group in the University of Cambridge Engineering Department's Electrical Engineering Division for information used in this chapter.

References

[1] R. A. Street, *Hydrogenated Amorphous Silicon*. Cambridge University Press, 1991.
[2] R. Fowler and L. Nordheim, Electron emission in intense electric fields, *Proc. Roy. Soc. Lond. Ser. A*, **119**, 173–181, 1928.

[3] C. Bandis and B. B. Pate, Photoelectric emission from negative-electron-affinity diamond (111) surfaces: Exciton breakup versus conduction-band emission, *Phys. Rev. B*, **52**, 12056, 1995.

[4] A. Ilie, A. Hart, A. Flewitt, J. Robertson, and W. Milne, Effect of work function and surface microstructure on field emission of tetrahedral amorphous carbon, *J. Appl. Phys.*, **88**, 6002–6010, 2000.

[5] L. Nilsson, O. Groening, C. Emmenegger, *et al.*, Scanning field emission from patterned carbon nanotube films, *Appl. Phys. Lett.*, **76**, 2071–2073, 2000.

[6] C. W. Tang and S. A. VanSlyke, Organic electroluminescent diodes, *Appl. Phys. Lett.*, **51**, 913–915, 1987.

[7] M. A. Baldo, D. F. O'Brien, Y. You, *et al.*, Highly efficient phosphorescent emission from organic electroluminescent devices, *Nature*, **395**, 151–154, 1998.

[8] R. H. Friend, R. W. Gymer, A. B. Holmes, *et al.*, Electroluminescence in conjugated polymers, *Nature*, **397**, 121–128, 1999.

[9] F. So, B. Krummacher, M. K. Mathai, D. Poplavskyy, S. A. Choulis, and V. E. Choong, Recent progress in solution processable organic light emitting devices, *J. Appl. Phys.*, **102**, 21, 2007.

[10] P. O. Anikeeva, C. F. Madigan, J. E. Halpert, M. G. Bawendi, and V. Bulovic, Electronic and excitonic processes in light-emitting devices based on organic materials and colloidal quantum dots, *Phys. Rev. B (Cond. Matt. and Mater. Phys.)*, **78**, 085434-1–085434-8, 08 2008.

8   Manufacturing and open innovation

T. Minshall, F. Livesey, L. Mortara, J. Napp, Y. Shi, and Y. Zhang

8.1   Introduction

This chapter discusses the changing ways by which value can be created and captured from nanotechnologies for future mobile devices. These issues are discussed from three perspectives. Firstly, attention is focused on how advanced technologies emerging from public and private laboratories can be transformed into products and services. Secondly, we look at how new, "open" models of innovation are emerging and discuss the opportunities and challenges this new approach presents for firms. Thirdly, we examine the changing ways in which manufacturing activities are coordinated and, in particular, the emergence of value creation networks as illustrated by changes in the mobile communications sector.

To allow discussion of these wide-ranging and interlinked activities, in this chapter we apply a broad definition of "manufacturing" as follows [1]:

Manufacturing is the full cycle of activity from understanding markets through design and production, to distribution and support, including after sales services and end of life management.

In this chapter, we will use the concept of the manufacturing value chain (see Figure 8.1) which allows us to consider the individual activities, the linkages between the activities, and also the manufacturing operation as a whole.

8.2   Commercialization of nanotechnologies

Commercialization of basic research outputs is typically a high-cost, long-timeframe, and high-risk activity. Such research outputs face numerous technological and market uncertainties. Basic research outputs are typified by being at a low level of "readiness" [2] but can be radical or incremental, disruptive or sustaining, and generic or specific in application.

As we have seen in previous chapters, nanotechnology (defined very broadly in this chapter as a field of science and technology in which one can control matter with at least one characteristic dimension in the range 1–100 nm) covers a wide range of technologies and associated application areas. Nanotechnologies can be considered to be radical (i.e.,

Nanotechnologies for Future Mobile Devices, eds. T. Ryhänen, M. A. Uusitalo, O. Ikkala, and A. Kärkkäinen. Published by Cambridge University Press. © Cambridge University Press 2010.

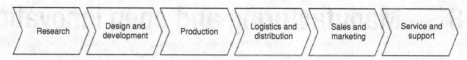

Figure 8.1 The generic manufacturing value chain [1].

a significantly higher performance over what is currently in use), generic (i.e., has a wide range of potential applications), and disruptive (i.e., offer a new value proposition) technologies.

Research has shown that the commercialization of such technologies can be particularly problematic [3] with the core problems including the need to resolve high levels of both market and technological uncertainties, and for substantial investment to be deployed over a long period of time [4, 5]. These uncertainties, coupled with the need for large-scale investment act as barriers to commercial adoption. To overcome these barriers, some have pointed to the need for the development of an integrated nanomanufacturing infrastructure, funded jointly by the private and public sector, to underpin the commercialization of nanotechnologies [6]. Such an infrastructure might include:

- geographically distributed nanomanufacturing research and nanofabrication user facilities;
- nanomanufacturing facilities that provide engagement with a range of stakeholders;
- support for small-business incubation and growth;
- efficient networking of nanomanufacturing facilities;
- standardization and documentation of nanofabrication procedures and process conditions, including libraries of components, processes, and models, as well as design tools.

Whether or not such programmes are implemented, the successful commercial exploitation of nanotechnologies will require organizations to consider new ways of innovating that bring together internal and external resources to create value. Taking a value chain view (as shown in Figure 8.2) allows consideration of the range of value creation and value capture options available to firms seeking to commercialize an advanced technology. However, it is extremely likely that value capture and creation relating to nanotechnologies will happen through organizations operating in collaboration rather than in isolation. In the following section we examine the global changes in processes by which new ideas are brought to market before examining the opportunities such changes bring to the commercialization of nanotechnologies for mobile devices.

8.3   New models of innovation

8.3.1   From closed to open innovation

In many industries and sectors, innovation is an increasingly distributed process, involving networks of geographically dispersed players with a variety of possible – and

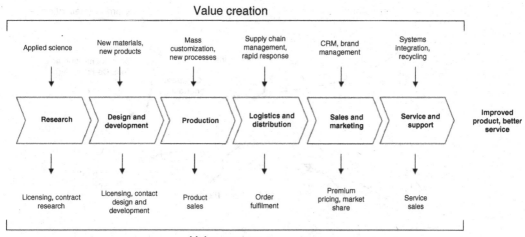

Figure 8.2 Opportunities for value creation and capture [1].

dynamic – value chain configurations. "Open innovation" (OI) is one term that has emerged to describe: "[..] the use of purposive inflows and outflows of knowledge to accelerate internal innovation, and expand the markets for external use of innovation, respectively" [7]. This is contrasted with the "closed" model of innovation where firms typically generate their own ideas which they then develop, produce, market, distribute, and support.

The trend towards OI is being driven by many factors which vary between firms and industries. There is also variation in the approaches taken by firms to the implementation of OI. However, there are some generic features of OI. With OI company boundaries become permeable and allow the matching and integration of resources between the company and external collaborators. In the closed innovation model, companies innovate relying on internal resources only.

Figure 8.3 presents a conceptual model of OI. The key feature of this diagram is that the company's boundaries become permeable. In a traditional closed innovation process all the invention, research, and development is kept secure and confidential within the company until the end-product is launched. With OI, the company can make use of external competencies (e.g., technologies) and even spin out by-products of its own innovation to outside organizations.

Although there will typically still be significant levels of internal activity at the research stage, there are also ideas and technologies that are developed outside, either collaboratively, or perhaps bought in. At the development phase, as research findings are narrowed down to viable projects, it may be advantageous to invest in externally developed intellectual property (IP) for certain technologies to advance these projects. This is OI inflow: the use of others' capability to innovate.

At the same time other IP that has emerged from the company's own research might be sold to other developers, either because it is of no strategic relevance to the company's

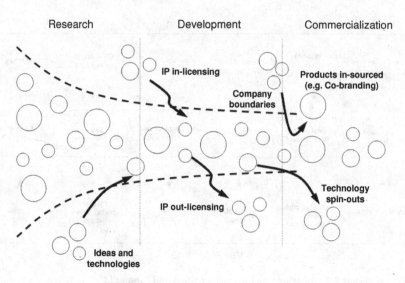

Research   Development   Commercialization

Figure 8.3 A conceptual model of OI [8, 9].

own current business, or because the company has no capacity or expertise to develop it itself. Alternatively, the company might see the opportunity to create technology spin-off companies to take on some of these projects. This is OI outflow: contributions made to others' innovation activities.

At the point of commercialization there will be core products that may have come through an entirely internal route from research to realization, or that may have had a variety of inputs from outside. Even at this stage, however, the OI company could choose to in-source market-ready products from outside. An example would be through co-branding where the OI company could use its established brand profile to sell a new product from another company that currently has no presence and credibility in the relevant market.

The successful implementation of OI requires consideration of a wide range of strategic and operational issues. In the following sections we review these issues relating to: (1) the different potential partners for OI, (2) the types of governance structures that can be used, and (3) the OI-related business processes.

8.3.2   Possible project partners

8.3.2.1   Customers and users

Customers and users have been known to be an important source of innovation, as users often generate new applications, techniques, products, and problem solutions out of personal need [10, 11]. Integrating customers' knowledge, needs, and specific requests into the new product development process may result in a range of possible benefits including:

- access to innovative resources of customers, resulting from experience with product application and from personal needs;

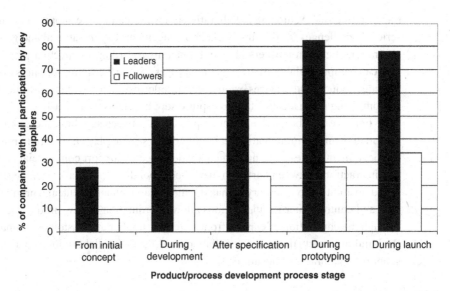

Figure 8.4 Key supplier involvement in the new product development (NPD) process [14].

- reduction of risk of product failure through targeting explicit customer needs;
- identification of future trends in markets through lead users, who are often starters and catalysts of new trends.

Challenges in integrating user innovations mainly occur in industries where the user is driven by direct personal need, but usually does not have any commercial interest. In these markets, users themselves test and retest their innovations and the transfer of problem-solving ideas from individual users to manufacturers can be difficult, as most of the generated knowledge is tacit [10, 12].

To meet this challenge, a number of different toolkits to support the involvement of customers have been developed [11, 13]. Two different types of those can be identified: those focusing on getting access to *need information* and those for getting access to *solution information*. Toolkits for getting access to *need information* provide users with a given solution space of the manufacturer for trial-and-error experimentation to facilitate the creation process. These toolkits are aimed at designing novel products and receiving immediate feedback on potential designs. Toolkits for getting access to *solution information* are aimed at more generic information about customer innovations. These toolkits provide a platform for interaction and communication to encourage thinking about problems and the transfer of ideas to the manufacturers.

8.3.2.2 Suppliers

Suppliers have been shown to be very important partners within the innovation process and are often involved throughout the whole new product development process, as illustrated in Figure 8.4.

Involving suppliers in the innovation process can help to share risks and costs of innovation, bringing in competencies, knowledge, and innovative ideas, and reducing transaction costs. Achieving collaborative innovation along the supply chain provides a series of challenges for the supply chain management. Organizational and geographical separations between partners [15] and concerns regarding intellectual property rights [14] can hamper collaborative innovation processes. Furthermore, suppliers often do not provide products and services exclusively for one customer and thus customers may end up competing against each other to capture supplier innovations [16].

To improve collaborative innovation processes between suppliers and customers, new information and communication technologies are seen to play an important role. Besides e-procurement and e-sourcing (which are designed to support commercial and contractual interactions between partners), new e-collaboration tools have been developed and tested. In addition to inventory management, demand management, and production planning and control, Internet applications allow coordination of new product development through the connection of experts from different supplier and customer companies [15]. By building a virtual space for collaboration, these e-collaboration tools can help to support collaborative innovation.

8.3.2.3 Start-up companies

For large companies start-ups can be attractive partners in the innovation process. Through alliances with entrepreneurial start-up firms, large companies can gain access to innovative, state-of-the-art technologies and to a new source of innovative capability. Furthermore, large firms can reduce their development risk by sharing technological uncertainty within a network of start-up companies. On the other hand, start-up companies benefit from cooperation with large companies through access to complementary assets and resources for manufacturing, distribution, and marketing as well as potential financial resources for a market launch of developed technologies [17–19]. Bringing these partners together in an OI environment is seen by some firms to be an attractive step, as both partners can clearly benefit from cooperation through access to complementary assets and knowledge as well as through collaborative learning.

However, research into the management of such "asymmetric" collaborations and alliances between large companies and start-ups has highlighted a number of challenges including issues of organizational cultural clashes, the ownership and use of intellectual property, expectation management, financial stability, and commercial maturity. To address these issues, management guides for partnerships between technology-based start-up firms and established companies have been developed to support the ability of advanced technologies to be commercialized through partnerships [20] but this remains a challenging management problem.

8.3.2.4 Universities

In addition to their traditional activities of researching and teaching, universities in many countries have been encouraged to build a "third mission" of supporting innovation through commercializing IP. Universities have the potential to contribute to companies' research and development processes directly and indirectly [21, 22]. Companies can therefore approach universities through different, often institutionalized, channels for collaboration: using intellectual property offered by technology transfer offices, building research collaborations with universities, sponsoring of PhD students, or through involvement of spin-off companies based on university research [8, 23–25].

Despite all the efforts supporting industry–university collaborations, various issues can still hamper cooperation between the partners. Issues in cooperation can be found in two main areas: IP management and cultural differences [26]. In the area of IP management, a lack of clarity over the ownership versus use of IP, and IP valuation seem to be key issues. Cultural differences also impede collaborations, as from an industrial point of view, universities are sometimes seen as bureaucratic, slow-moving, and risk-averse [26].

8.3.2.5   Public research institutions (PRIs)

As well as universities, other PRIs can be important sources of innovation. These institutes are usually established by federal or local governments to support technology transfer to private business. Like universities, these research institutions can offer access to state-of-the-art technologies and thus can be valuable partners in the innovation process. Even though their research focus often differs from that of universities and is more oriented towards practical application, strategic and operational issues at these research institutes can also pose problems for collaboration [27]. PRIs are sometimes viewed as sitting in a gray zone between commercial and academic values. For these institutions, external funding based on industry collaboration and IP licensing may be more important than it is for universities. As such, conflicts may arise when defining collaboration with industry related to IP ownership.

PRIs can represent an important link between the academic research (discoveries) and industrial research and development (applied research, inventions, products), and can provide the capabilities essential to ramp up technologies towards industrial scale.

8.3.2.6   Intermediaries

In the era of OI, intermediaries can play an increasing role in connecting buyers and sellers in the market for innovation and ideas, as they increase the operational effectiveness of these markets and expose supply and demand to a wider and more sophisticated group of potential partners [28, 29]. Depending on their innovation strategy, companies can access intermediaries with different characteristics, and intermediaries can themselves be grouped into three different types: invention capitalists, innovation capitalists, and venture capitalists. They vary in objectives, function, ways of adding value, core competencies, assumed risks, willingness for capital investment, ownership of IP rights, and relationship with the client company. Companies need to choose a balanced approach to innovation sourcing, considering both the industry and market profile as well as the internal company structure [29]. The growth in on-line resources to support innovation has been one noticeable trend. Organizations such as www.innocentive.com and www.ninesigma.com provide companies with routes to both markets and resources to support open innovation.

8.3.3   Forms of collaboration

Collaboration with the partners described in the previous section can be enabled via a range of possible governance structures [30]. The following subsections give a brief overview of these different forms of collaboration and their importance for an open approach to innovation.

8.3.3.1   Joint venture

Joint ventures are a type of organizational structure formed "... when two or more firms pool a portion of their resources within a common legal organization" [31]. Joint ventures usually require equity investments and involve the foundation of a new business unit, jointly run by the partner firms. Due to their structure, joint ventures are comparatively inflexible to changes and the exchange of know-how and skills is limited to a small number of involved project partners, usually with a long time horizon and a high interdependency. Joint ventures represent a comparatively closed degree of organizational openness, but can be used for technology acquisition as well as for exploitation.

8.3.3.2   Strategic alliances

The term "strategic alliance" is used to specify a range of interorganizational relationships in which autonomy is maintained by the partners, but they are bilaterally dependent to a nontrivial degree [32]. The dependence can be via common ownership, contractual arrangements, or in some informal way. By providing a mechanism to access a wider range of tangible and intangible resources than are available within one organization [33] as well as by spreading risk, alliances would seem to be an effective way to enhance innovation.

However, benefits may not be realized owing to the tension that can be seen to exist between the dynamics of innovation and the logic required for alliances to be a success. Innovation stems from creativity, uncertainty, and risk-taking; alliances need to emphasize clarity and explicitness [34]. Despite this incompatibility, strategic alliances are commonly observed around emerging technologies, and certain forms of alliance have shown themselves to be more suited to such environments.

Equity forms of alliance are more likely to stress issues of control. Specifically, there may be more emphasis on setting targets, measuring against these targets, and taking corrective actions. More flexible, evolution-oriented, nonequity forms of alliance may promote day-to-day cooperation more than equity forms. Nonequity alliances may also better promote reciprocal information exchange and the development of a common language than do joint ventures and partial equity alliances [30, 35, 36].

The ability to form and manage strategic alliances can be considered as a distinctive organizational capability that the firm may develop which provides a mechanism for accessing and reconfiguring resources in response to changing conditions. This organizational capability is one that can be developed and enhanced to allow a firm to extend the boundaries of its accessible resources.

8.3.3.3 Licensing

Licensing generally plays a very important role in implementing an OI strategy and there has been a substantial increase in the technology licensing activities of firms, matching the observed trend towards OI [37]. For an open approach to innovation, two directions of licensing have to be distinguished: in-licensing and out-licensing. In-licensing allows the acquisition of technology developed by external parties. In contrast, out-licensing is part of the technology selling process. It offers a low-risk method to commercialize technology which has been developed within the company, but which has not been commercialized within its own products [38].

The design of contracts in both in- and out-licensing varies regarding the degree of control that the licensor can retain and the licensee can gain over the technology. Options to profit from licensing are, e.g., charging a fixed one-time fee, a per unit royalty, tied purchases of inputs from the licensor, or provisions regarding cross-licensing of patents. Additionally, restrictions on the markets in which the licensee may sell goods and the possibility of claiming equity shares might minimize the risk associated with licensing for the licensor, but limit the freedom of the licensee.

In general, the main purpose of in-licensing can be seen as fast access to technology at a low cost [39], whereas out-licensing aims at generating revenues from unused technologies, but also at generating strategic advantages through the setting of standards, ensuring technology leadership, and enhancement of reputation [37].

Depending on the design of the contract, in- and out-licensing activities can help to implement an open approach to innovation, as they allow the involvement of many partners and offer a flexible way of technology transfer through firm boundaries in both directions.

8.3.3.4 Spin-outs

Spinning out ventures offers the opportunity to create and capture value from technology developed inside an organization – be it a firm, a university, or a research institute – through external channels. A spin-out offers a way to externalize unused technology and thus is part of the technology exploitation strategy [40].

Within the spin-out process, access to human, social, financial, physical, technological, and organizational resources can be identified as key. Through initial as well as continued support of the parent organization, spin-outs can gain access to these resources, while the parent organization maintains influence on the development of the spin-out. If linked to corporate venturing activities, parent organizations can benefit from spin-outs financially as well as strategically.

8.3.3.5 Informal relationships

Within an open approach to innovation, informal, noncontractual collaborative activities between companies and external parties can play an important role. Collaborative activities may include informal impromptu meetings, virtual or in-person brainstorming sessions, and general unrestricted exchange of ideas and information. Ways to support these activities are manifold. Exchange of information can be in person at organized meetings (especially encouraged within regional networks and clusters), via e-mail,

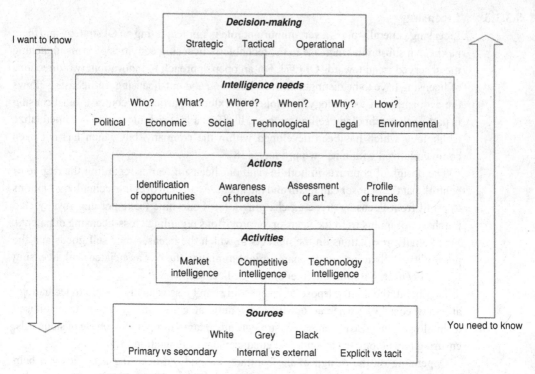

Figure 8.5 The intelligence framework [43].

phone, or through online- or video-conferencing [41]. Universities can play an important role in the setup of informal relationships through the provision of "public spaces" that encourage social interaction between industry and academia [42].

8.3.4   Processes related to OI

8.3.4.1   Technology intelligence

Companies who want to implement an open approach to innovation face the challenge of keeping track of external technology opportunities, risks, and threats. Technology intelligence activities provide companies with the capability to monitor, capture, and deliver information about technological development. Technology intelligence can be defined as "... the capture and delivery of technological information as part of the process whereby an organization develops and awareness of technology threats and opportunities" [43].

As illustrated in Figure 8.5, intelligence is used to keep decision makers informed on strategic, tactical, and operational matters. Information is needed on many issues surrounding a technology such as its environmental, social, and legal impact, as well as the technical details. Decision makers are the intelligence consumers, whereas the intelligence activities are performed by those who search for information (i.e., intelligence brokers) from different sources. The prompt for the intelligence activity may come from

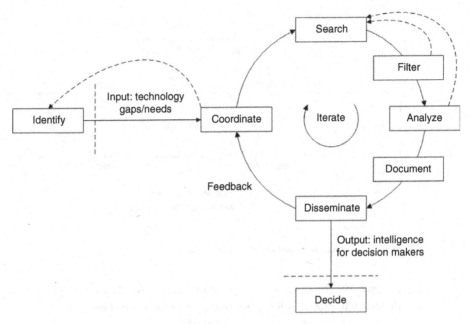

Figure 8.6 The operating cycle for technology intelligence [43].

the decision makers' need to know, but is also promoted by the intelligence brokers when something interesting is identified. The company "actions" intelligence for many purposes such as the identification of opportunities, the generation of threat awareness, the assessment of the state-of-the-art, and the profiling of trends [43]. Often technology, market, and competitor information are collected separately. It is important to encourage communication between the three areas as both market and competitor intelligence can contribute valuable information about new technologies.

Within a firm, these activities can be organized through different intelligence structures: actions could be undertaken by a dedicated central unit, included in a central group function, decentralized to operational divisions, or diffused throughout the company [44]. The intelligence activity itself can be coordinated structurally through hierarchical delegation of tasks, informally through stimulation of autonomous information gathering behavior, or as a hybrid by delegating tasks through projects of limited duration, focused on single topics [45].

Figure 8.6 shows the operating cycle for intelligence [43]. Following this iterative cycle within technology intelligence processes can help to assure the quality of gathered information regarding well-defined scope, accuracy of collected data, robustness of data analysis, and objective findings and conclusions [43, 46].

8.3.4.2 IP management

As ideas and technologies become increasingly shared within OI projects between partners, management and protection of intellectual property rights (IPR) will be increasingly important [8, 47]. As important as it is to share proprietary intellectual assets

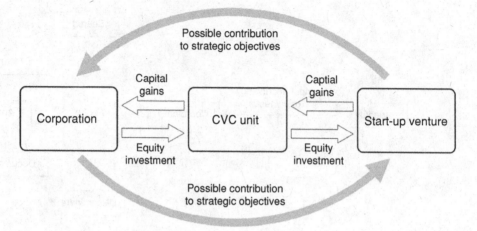

Figure 8.7 Typical structure of corporate venturing.

to meet the objectives of collaborations within the innovation process, it is equally important to protect IPR against improper or unintended use [48].

Even though agreements might not offer complete protection against improper use of any intellectual asset, they serve two important functions: Firstly, they protect the patented IPR and, secondly, they raise the awareness of the involved employees regarding the ownership of intellectual assets and shared information [48].

To allow the most effective protection of intellectual assets, the creation of specific, dedicated organizations within the companies is seen to be crucial [47, 48]. To maximize the value of their intellectual asset portfolio, tracking of existing IP and of collaborative agreements can be arranged through these organizations. Intangible asset management (IAM) systems and commercial tracking tools can help to "... find, share, reuse, distribute and archive information, and monitor legal status, licensing and royalty tracking" [47]. Additionally, IP management can help to identify unused, but valuable intellectual assets and to commercialize them through out-licensing [47].

8.3.4.3   Corporate venturing

Corporate venturing can be generally defined as "... any form of business development activity that leverages corporate mass to create competitive advantage in core business, or to grow related strategic markets" [49]. Especially in large companies, corporate venturing units are created for these activities and endowed with a certain stock of capital or investment budget. In these cases, the corporate venturing units usually act as intermediaries between new business ventures and the parent companies and invest the companies' funds in start-up firms [50]. The typical structure of corporate venturing is illustrated in Figure 8.7.

Corporate venturing activities are conducted for various different reasons, such as the development of an ecosystem around a particular commercial opportunity, growth of existing business, entering new business (within the "white space"), reduction of risks, or generating cash from internal ventures [49, 51]. Furthermore, corporate venturing units can be used as a tool for technology and marketing intelligence, as they can provide entrepreneurial insights regarding emerging technologies and markets [49, 51, 52].

To achieve these aims, different investment opportunities exist. Internal corporate venturing focuses on generating routes for technology exploitation. Such internal corporate venturing is aimed at spinning-out interesting initiatives with high growth potential which have been initiated within the company, but do not fit the original strategic plan of the company, or which are perceived as too risky to realize within the product portfolio. These internal venturing activities can help to rejuvenate a firm's business portfolio and to increase research and development efficiency through commercializing unused intellectual assets [50, 51, 53].

On the other hand, external corporate venturing activities can aim at gaining access to external investment opportunties, which are of strategic interest for the company. By investments in external ventures – either passive though investment in venture capital funds or active as investor with direct involvement in management activities – general monitoring of, and access, to new technologies and applications can be achieved [50, 51, 53].

8.3.5 OI and nanotechnologies

So what is the link between OI and the commercial application of nanotechnologies? Earlier in this chapter we highlighted some of the challenges in the commercialization of a radical, generic, disruptive technology, characteristics of many nanotechnologies. These challenges included the need to resolve high levels of both market and technological uncertainties, and for substantial investment to be deployed over a long period of time [5]. Given this situation, it is extremely unlikely that any one organization is going to be in possession of sufficient breadth of resources and patient capital to allow a closed approach to bringing nanotechnologies to market. It is recognized that nanotechnology commercialization relies upon an awareness of the business ecosystem and the formation of effective alliance partnerships with a diverse range of organizations. This would seem to point to the need for organizations seeking to create value from nanotechnologies to develop the organizational culture, structures, and skills required to effectively implement OI.

As discussed earlier, a major challenge facing the successful implementation of OI in any industry sector is the effective management of IPR. However, nanotechnologies have particular IPR-related issues associated with them that can exacerbate implementation problems. The successful exploitation of nanotechnologies may well rely upon the combination of IP from multiple sources including internal research and development, suppliers, alliance partners, universities, and public research institutes. Nanotechnologies are also reaching a level of maturity where the emphasis for key firms has shifted from basic research to applications, while at the same time using IPR to generate licensing income. This shift in focus for the leading firms and the number of organizations

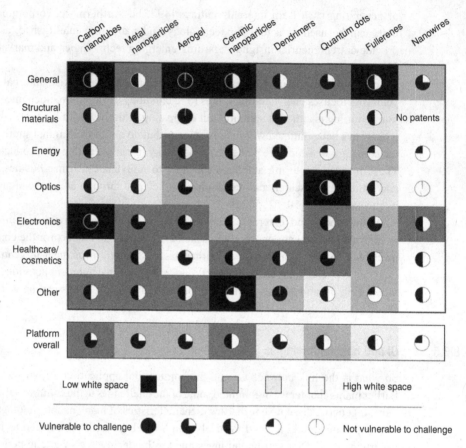

Figure 8.8 IPR entanglement and areas of "white space" for selected nanotechnologies and application areas. *Source:* Lux Research [54].

continuing to develop and exploit IP are increasing the likelihood of IPR entanglement and the reduction of areas of "white space" [54] (see Figure 8.8).

Collaborative models of IP generation and commercialization involving academic and commercial partners can be observed around nanotechnologies. These can take the form of either bilateral collaborations or multiparty collaborations. An example of the former can be seen in Nokia Research Centre's collaboration with a research group at the University of Cambridge Engineering Department to develop nano-enabled solutions to increase energy density and speed up charge/discharge cycles in energy storage systems, and also to investigate the potential for novel, flexible form-factors in batteries and supercapacitors [55].

An example of a multiparty collaboration can be seen in the work of the Cambridge Integrated Knowledge Centre (CIKC)(www.cikc.co.uk). The CIKC is focused on research, training, and commercialization relating to molecular and macromolecular materials. By bringing together multiple university departments along with corporate partners ranging from multinational corporations through to small and medium-sized enterprises (SMEs), the CIKC is seeking to accelerate the research, initial commercialization, and ramp up of molecular and macromolecular materials.

The vision presented in Chapter 1 is one of near-ubiquitous access to information and virtual resources. OI provides an approach for leveraging such access to enable value creation and capture. But successful exploitation is much more than just the initial entry into the market: successful nanotechnology innovators must also be able to access the full range of manufacturing capabilities to ramp up operations globally and sustainably. In the next section we review the changes impacting elements of the manufacturing value chain and how their reconfiguration offers opportunities for improving the effectiveness of the commercialization of advanced technologies such as nanotechnology.

8.4 New configurations of manufacturing capabilities

For industries and sectors with clearly defined value chains firms can seek either to manage an integrated set of activities directly under their ownership, to specialize in selected activities along the value chain, or to act as an integrator of the activities of other firms along the value chain. The mobile communications sector illustrates the way in which manufacturing capabilities can be reconfigured to address changes in technologies, production capabilities, and markets.

8.4.1 Manufacturing and mobile communications

8.4.1.1 From first generation (1G) to third generation (3G)

The mobile communications industry has expanded from serving performance-orientated business users to serving a price-sensitive and functionality-orientated mass market. The industry value chain has evolved from a simple linear system dominated by a small number of highly integrated and independent big companies (see Figure 8.9) to a complex network consisting of numerous specialized and interdependent companies of different sizes (see Figure 8.10) [56].

The mobile communications industry value chain has become increasingly complex since the 1990s. There has been a significant trend for the specialization, modularization, and segmentation of handset manufacturing (see Figure 8.10). Many companies have successfully entered the industry by focusing on a small section of the value chain/value network, and reinforced their leading position by extending their businesses up-stream and/or down-stream. Vertically integrated manufacturing systems are not the only choice for handset manufacturers. They can simply collaborate with a set of suppliers to develop and produce their mobile phones.

Innovative business models have also emerged by combining the value chain segments in different ways. For example, a service provider may work with some specialized suppliers of software platforms and mobile phone components to develop a new model, and engage a contract manufacturer to produce that model with their own brand.

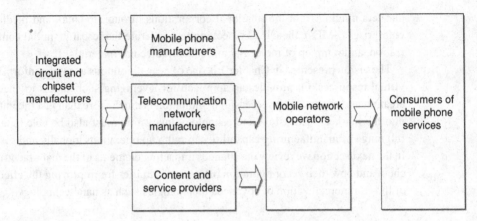

Figure 8.9 A simplified view of key players in the mobile communication value chain [56].

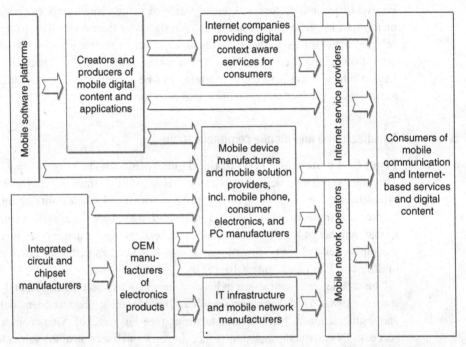

Figure 8.10 Key players in the mobile communication value chain today. The industry is characterized by the convergence of telecommunications, consumer electronics, and IT sectors [56].

8.4.1.2   The emergence of global value chain networks

The value chain in the mobile communications industry now consists of a broad range of interrelated value-adding activities from idea generation and research towards design, development, component fabrication, system assembly, testing, marketing, distribution, and service. Evolving with the environment, the value chain has migrated to

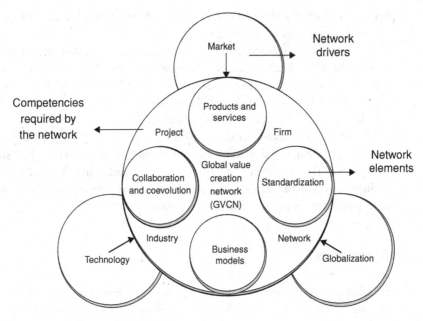

Figure 8.11 Key elements of a GVCN.

a global value creation network (GVCN) (see Figure 8.11). A GVCN represents an emerging manufacturing system which is much more complex than a linear value chain. It is a relationship-based value network including product level development and design, sector level standardization, individual firm level business model innovation, and wider ecosystem development for OI. The main building blocks of a GVCN include products/services, standardization, business models, and collaboration/coevolution.

Products or services in the mobile communications sector have unique characteristics. For example, mobile terminals are physical products that integrate complex technology inside with personal preference outside; the usage of the mobile phone depends on not only the terminals but also the service providers' offers and their further networks. These combinations lead to unpredictable product changes. From a production and supply network perspective, product platforms and modularity are critical as mechanisms to cope with volatile changes and demand for production efficiency.

Standardization in the mobile communications industry has two types of implications. At the physical system level, standardization derives from the tradition of the electronics industry to adopt standards and protocols in a wider range of components and systems. This benefits product development and system integration, especially in the case of radical and fragmented changes. At the technology generation level, the industry has developed different generations of telecommunications protocol standards in order to provide a special way to "integrate" radical innovations and exploit both technology and market developments. Successive generations of standards of telecommunications (i.e. 1G, 2G, and 3G) have very successfully underpinned spectacular industry growth.

Standards at both the physical systems level and the technological level have thus played a key role in the success of this industry.

As well as product and process innovations, companies in the mobile communications industry have widely adopted business model innovation throughout the evolution of the industry. Mobile communication companies have had to position or reposition their core business and capabilities within collaborative interfirm networks, and to learn to coordinate their activities along the whole value chain. There has been a significant trend towards more mobile communication companies adopting interfirm collaborative models rather than classical vertically integrated models to achieve their strategic goal and growth. The mobile communications sector is adopting many of the OI practices outlined earlier in this chapter to cope with the rapid pace and radical nature of change in technologies, markets, and industry structure.

8.5   Conclusions

So what do these contextual changes in manufacturing and innovation mean for the commercialization of nanotechnologies in general and specifically for their application in future mobile devices? As previous chapters in this book have shown, there are numerous ways in which nanotechnologies have the potential to impact upon future mobile devices. But nanotechnology is not just "any" radical, generic, disruptive technology with an associated set of challenges affecting its commercialization. It is a technology that has the potential to change the very way we innovate, the way we fabricate devices, and the way in which we organize manufacturing.

This chapter has highlighted the challenges facing nanotechnology-based innovations that are common to any radical, generic, and disruptive innovations. Key among these are the need to cope with high levels of technology and market uncertainty, and the need for substantial and patient capital to cope with the typically long time scales involved in moving from laboratory to market. Coupled with these challenges has been the growing recognition that successful nanotechnology commercialization will require a collaborative, open approach to innovation. Such an open approach helps companies cope with the market and technological uncertainties by sharing the risks and costs, but also presents companies with numerous management challenges and increases the potential for disputes around the ownership of IPR and associated value.

This chapter has also presented the mobile communications industry as an example of an industry that has morphed from a series of largely vertically integrated players towards an industry characterized by the convergence of the telecommunications, consumer electronics, and IT sectors, and the development of value creation networks made up of a range of organizations from start-ups to multinationals. This convergence is also leading to an increasing emphasis being placed upon the capturing of value from content rather than devices for many of the industry players.

In conclusion, the success of the mobile communications industry in embracing an open, networked approach to value creation and capture may position it well to benefit from the potential presented by the emergence of nanotechnologies.

References

[1] M. Gregory, P. Hanson, A. J. van Bochoven, and F. Livesey, *Making the Most of Production.* University of Cambridge Institute for Manufacturing, 2003.

[2] J. C. Mankins, *Technology Readiness Levels: A White Paper.* 1995, NASA. From: http://www.hq.nasa.gov/office/codeq/trl/trl.pdf.

[3] J. D. Linton and S. T. Walsh, A theory of innovation for process-based innovations such as nanotechnology, *Technol. Forecasting Social Change,* **75**, 583–594, 2008.

[4] S. Lubik and E. W. Garnsey, Commercializing nanotechnology innovations from university spin-out companies, *Nanotechnol. Perceptions,* **4**, 225–238, 2008.

[5] E. Maine and E. W. Garnsey, Commercializing generic technology: The case of advanced materials, *Res. Policy,* **35**, 375–393, 2006.

[6] J. Chen, H. Doumanidis, K. Lyons, J. Murday, and M. C. Roco, *Manufacturing at the nanoscale: National Nanotechnology Initiative Workshop Report.* National Science and Technology Council, Committee on Technology, Subcommittee on Nanoscale Science, Engineering and Technology, 2007.

[7] H. Chesbrough, W. Vanhaverbeke, and J. West, eds. *Open Innovation: Researching a New Paradigm,* Oxford University Press, 2006.

[8] H. Chesbrough, *Open Innovation: The New Imperative for Creating and Profiting from Technology,* Harvard Business School Press, 2003.

[9] M. Docherty, *Primer on 'Open Innovation': Principles and Practice,* pp. 13–17, Vision (Product Development and Management Association), 2006.

[10] E. von Hippel, *The Sources of Innovation,* Oxford University Press, 1988.

[11] E. von Hippel, *Democratizing Innovation,* The MIT Press, 2005.

[12] C. Hienerth, The commercialization of user innovations: the development of the rodeo kayak industry, *R&D Management,* **36**, no. 6, 273–294, 2006.

[13] F. T. Piller and D. Walcher, Toolkits for idea competitions: a novel method to integrate users in new product development, *R&D Management,* **36**, no. 3, 307–318, 2006.

[14] J. D. Blascovich and W. Markham, How procurement excellence creates value, *Supply Chain Management Rev.,* **9**, no. 5, 44–52, 2005.

[15] R. L. Chapman and M. Corso, From continuous improvement to collaborative innovation: the next challenge in supply chain management, *Production Planning & Control,* **16**, no. 4, 339–344, 2005.

[16] Procurement Strategy Council, *First Among Equals – Using Customer of Choice Strategies to Capture Supplier Innovation,* 2006.

[17] S. A. Alvarez and J. B. Barney, How entrepreneurial firms can benefit from alliances with large partners, *Acad. Management Executive,* **15**, no. 1, 139–148, 2001.

[18] A. De Meyer, Using Strategic partnerships to create a sustainable competitive position for high tech start-up firms, *R&D Management,* **29**, no. 4, 323–328, 1999.

[19] T. H. W. Minshall, L. Mortara, and J. J. Napp. Implementing open innovation: challenges in linking strategic and operational factors for HTSFs working with large firms. In *Proceedings of the 15th High Tech Small Firms Conference (2007),* Manchester Business School, 2007.

[20] T. H. W. Minshall, L. Mortara, S. Elia, and D. Probert, Development of practitioner guidelines for partnerships between start-ups and large firms, *J. Manufacturing Tech. Management,* **19**: 3, 391–406, 2008.

[21] W. M. Cohen, R. R. Nelson, and J. Walsh, Links and impacts: the influence of public research on Industrial *R&D, Management Sci.,* **48**, no. 1, 1–23, 2002.

[22] K. Laursen and A. Salter, Searching high and low: what type of firms use universities as a source of innovation?, *Research Policy*, **33**, no. 8, 1201–1215, 2004.

[23] M. Abreau, V. Grinevich, A. Hughes, M. Kitson, and P. Ternouth, *Universities, Business and Knowledge Exchange*. Council for Industry and Higher Education, University of Cambridge Centre for Business Research, 2008.

[24] H. Chesbrough, *Open Business Models: How to Thrive in the New Innovation Landscape*, Harvard Business School Press, 2006.

[25] J. de Wit, B. Dankbaar, and G. Vissers, Open innovation: the new way of knowledge transfer?, *J. Business Chemistry*, **4**, no. 1, 11–19, 2007.

[26] R. Lambert, *Lambert Review of Business-University Collaboration: Final Report*. HM Treasury, www.lambertreview.org.uk, 2003.

[27] I. Drejer and B. H. Jørgensen, The dynamic creation of knowledge: analysing public private collaborations. *Technovation*, **25**, no. 2, 83–94, 2005.

[28] J. S. Gans and S. Stern, The product market and the market for "ideas": commercialization strategies for technology entrepreneurs, *Res. Policy*, **32**: 2, 333–350, 2003.

[29] S. Nambisan and M. Sawhney, A buyer's guide to the innovation bazaar. *Harvard Business Rev.*, **85**: 6, 109–118, 2007.

[30] V. van de Vrande, C. Lemmens, and W. Vanhaverbeke, Choosing governance modes for external technology sourcing, *R&D Management*, **36**, no. 3, 347–363, 2006.

[31] B. Kogut, Joint ventures: Theoretical and empirical perspectives, *Strategic Management J.*, **9**, 319–332, 1988.

[32] O. E. Williamson, Comparative economic organisation: The analysis of discrete structural alternatives, *Administrative Sci. Quart.*, **36**, 269–296, 1991.

[33] D. Leonard-Barton, *Wellsprings of Knowledge: Building and Sustaining Sources of Innovation*, Harvard Business School Press, 1995.

[34] F. Bidault and T. Cummings, Innovating through alliances: experiences and limitations, *R&D Management*, **24**, no. 1, 1994.

[35] F. Chiaromonte, Open innovation through alliances and partnership: theory and practice, *Int. J. Technol. Management*, **33**, no. 2/3, 111–114, 2006.

[36] R. N. Osborn and J. Hagedoorn, The institutionalization and evolutionary dynamics of interorganizational alliances and networks. *Acad. Management J.*, **40**, no. 2, 261–278, 1997.

[37] U. Lichtenthaler, Corporate technology out-licensing: motives and scope, *World Patent Information*, **29**: 2, 117–121, 2007.

[38] S. Vishwasrao, Royalties vs. fees: how do firms pay for foreign technology?, *Intl. J. Ind. Organization*, **25**, no. 4, 741–759, 2007.

[39] C. Tomkins, Interdependencies, trust and information in relationships, alliances and networks, *Accounting, Organisations and Society*, **26**, no. 2, 161–191, 2001.

[40] R. Agarwal, R. Echambadi, A. M. Franco, and M. B. Sarkar, Knowledge transfer through inheritance: spin-out generation, development, and survival, *Acad. Management J.*, **47**, no. 4, 501–522, 2004.

[41] M. Swink, Building collaborative innovation capability, *Research Technology Management*, **49**, no. 2, 37–47, 2006.

[42] A. Cosh, A. Hughes, and R. K. Lester, *UK plc: Just How Innovative Are We?*, Cambridge MIT Institute, www.cambridge-mit.org, 2006.

[43] C. I. V. Kerr, L. Mortara, R. Phaal, and D. R. Probert, A conceptual model for technology intelligence, *Intl. J. Technol. Intelligence and Planning*, **2**, no. 1, 73–93, 2006.

[44] J. J. Quinn, How companies keep abreast of technological change, *Long Range Planning*, **18**, no. 2, 69–76, 1985.

[45] E. Lichtenthaler, Technological change and the technology intelligence process: a case study, *J. Eng. Technol. Management*, **21**, no. 4, 313–348, 2004.

[46] R. Schwartz and J. Mayne, Assuring the quality of evaluative information: theory and practice. *Evaluation and Program Planning*, **28**, no. 1, 1–14, 2005.

[47] J. Hogan, Open innovation or open house: how to prototect your most valuable assets, *Medical Device Technology*, **16**, no. 3, 30–31, 2005.

[48] G. Slowinski, E. Hummel, and R. Kumpf, Protecting know-how and trade secrets in collaborative R&D relationships, *Research Technol. Management*, **49**, no. 4, 30–38, 2006.

[49] M. O. L. Collins, Venturing a solution. *British J. Administrative Management*, **49**, 26–27, 2005.

[50] H. Ernst, P. Witt, and G. Brachtendorf, Corporate venture capital as a strategy for external innovation: an exploratory empirical study, *R&D Management*, **35**, no. 3, 233–242, 2005.

[51] S. K. Markham, S. T. Gentry, D. Hume, *et al.*, Strategies and tactics for external corporate venturing, *Research Technol. Management*, **48**, no. 2, 49–59, 2005.

[52] W. Vanhaverbeke and N. Peeters, Embracing innovation as strategy: corporate venturing, competence building and corporate strategy making. *Creativity and Innovation Management*, **14**, no. 3, 2005.

[53] V. van de Vrande, C. Lemmens, and W. Vanhaverbeke, Choosing governance modes for external technology sourcing, *R&D Management*, **36**, no. 3, 247–363, 2006.

[54] Lux Research, *The Nanotech Report Investment Overview and Market Research for Nanotechnology*, Lux Research Inc., 2007.

[55] Nokia, *Open Threads: Nokia Open Innovation Newsletter*, Nokia Research Centre, 2009.

[56] T. Ryhänen, From personal communication between Tapani Ryhänen (Nokia Research Center) and the authors. 2009.

9 Seeing beyond the hype: what the Internet teaches us about the development of nanotechnology

T. Crawley, L. Juvonen, and P. Koponen

9.1 Introduction

In the following, we introduce two frameworks for the analysis of technology development: the hype cycle and the S-curve. These are used to make a comparison between two general-purpose technologies: the development of the Internet during the last decade, and the current state of nanotechnology development. The analysis provides evidence that we are at the start of a period in which nanotechnology will begin to have a significant economic impact, a period labelled "Nanotech 1.0."

9.2 Perception and performance

The general level of interest toward a new technology typically evolves through a number of phases ranging from overenthusiasm (hype) to subsequent disappointment. This is described by the hype cycle (see Figure 9.1), first developed by Gartner in 1995 [1]. The hype cycle employs some acute insights into human behavior to characterize how perception relates to the actual maturity of a new technology.

The hype cycle begins with a breakthrough phase that may be the first description of a new technology, the first research results suggesting the existence of a phenomenon, or the first generation of a product. This technology trigger catalyzes interest which increases rapidly and soon the publicity grows out of proportion to the true state of development. This overenthusiasm culminates in a so-called peak of inflated expectations, at which point it is realized that the technology is unlikely to be able to satisfy these high expectations in the short term. Interest falls rapidly and the technology quickly becomes unfashionable, entering a state called the trough of disillusionment.

Yet whilst the general perception is that the technology has failed to meet the expectations, in reality development continues with further efforts to understand the benefits and practical applications of the technology. Once this comes to wider attention, the technology enters a period in which perception and reality are largely in line: the slope of enlightenment. Finally the technology reaches a period called the plateau of productivity where its utility is high and, despite not necessarily generating high public excitement, its benefits are demonstrated and accepted.

Nanotechnologies for Future Mobile Devices, eds. T. Ryhänen, M. A. Uusitalo, O. Ikkala, and A. Kärkkäinen. Published by Cambridge University Press. © Cambridge University Press 2010.

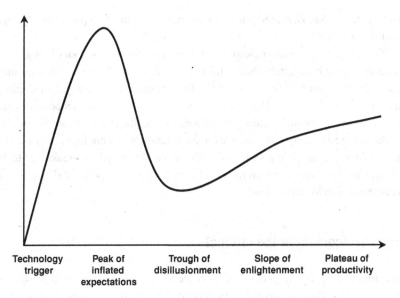

Figure 9.1 The hype cycle, adapted from [1].

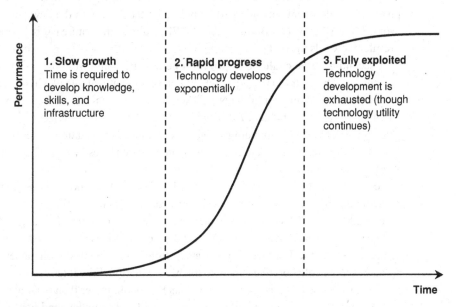

Figure 9.2 The S-curve.

Looking beneath the hype, the development in the actual performance of a new technology is commonly described by an S-curve (see Figure 9.2). The first phase of development represents a period in which fundamental knowledge and understanding of a technology are established. This may include research, patenting activity, development of tools and instruments, and standardization. Because this phase primarily focuses on what a technology is, rather than what it can do, its impact on potential application areas is still very limited.

Phase two sees the technology evolve rapidly as a common stock of knowledge and processes have been established. This is also the period during which a technology generates the most interest from industry and the financial community. In the final phase, the development of the technology slows. This may be as a consequence of insurmountable technical or economic barriers, or as a result of displacement by another technology.

As nanotechnology encompasses a wide range of technologies, its development is more likely to resemble a series of such curves, separated by breaks or "technological discontinuities" during which the development of one technology slows, just as another is an early development phase.

9.3   The development of the Internet

The Internet, a previous general-purpose technology, provides a good illustration of these two concepts. The Internet has experienced at least two S-curves, the first of which was a phase of development which began in the late 1980s and culminated with widespread adoption of the first commercial web browser, Netscape, in the mid–late 1990s. The phase of development was initiated by public investment, with the development of the World Wide Web by Tim Berners-Lee at CERN, although venture capital became more prevalent towards the end of this period.

A second S-curve could be drawn for a period stretching from the late 1990s to the present day. This is also a period which gives a very clear example of a hype cycle, if one considers the development of the NASDAQ share index in the period 1999–present as a suitable proxy for the perception of the Internet. The period 1999–2000 saw a massive share price bubble, corresponding to the peak of inflated expectations. This was followed by a dramatic crash, in which many companies' share prices were slashed, whilst other ceased to exist altogether.

Yet in 2003, the NASDAQ commenced a four-year period of consistent growth, appreciating by 80% over the period and resembling closely the hype cycle's slope of enlightenment. This period coincides with the development of a range of services categorized by Tim O'Reilly [2] as Web 2.0. These services represented a fundamental reorientation of the Internet, moving away from a unidirectional client/server model towards a more collaborative "many to many" orientation. Examples of this shift include the increasing use of social networks such as Facebook, rather than personal webpages.

It is worthwhile to look behind this perceptual level to understand how the Internet industry was actually developing. Firstly, the number of Internet users was increasing at an impressive rate, from 147 million users in 1998 to 1.32 billion in 2007. Whilst the share price bubble took a destructive toll on shareholders, it had also provided the funding for an expansion of the infrastructure required for the Internet to flourish: the server farms, fiber-optic trunk lines, and broadband connections that were introduced in much of the developed world. Behind the hype, investment was driving the development of these new services, though their benefits would not be felt until later.

9.4   Reaching maturity: Nanotech 1.0

Since the year 2000, nanotechnology has received large-scale public investment which has been used to fund infrastructure and stimulate research activity. However, it appears that in terms of interest from the public and the financial community, nanotechnology is languishing in the trough of disillusionment. There is a sense that nanotechnology has somehow failed to deliver, and even that the risks of the technology are outweighing its benefits. In the following, we argue that in actuality, nanotechnology is poised on the cusp of an era in which it will have a significant economic impact, a period labelled Nanotech 1.0.

There are few better measures of "hype" than Google's own search data. Whilst the company does not release the exact number of searches for a given term, the company's Google Trends utility does indicate the rate of increase or decline. Searches for the term "nanotechnology" carried out by Google users in the USA have shown a steady decline since 2004. Approximately 60% fewer monthly searches for the term were carried out in 2009 than at the start of 2004 [3].

There are also a number of indexes which have been created to track nanotechnology stocks. One of these, the Lux Powershares index, has underperformed relative to the Standard & Poor 500 since it was established in late 2005. The index has lost 47.02% of its value, against a 24.28% drop for the Standard & Poor 500 as a whole. Whilst this indicates that the financial community takes a rather dim view of nanotechnology, it also underlines the point that there has not been a point at which private investment was driving nanotechnology hype.

The measures suggest that one would currently locate nanotechnology on the hype cycle as being in the Trough of Disillusionment. This is supported by discussions with members of the nanotechnology community, who, whilst being wholly committed to the technology, detect a lack of interest from the wider public.

However, as with the Internet, a general mood of disinterest masks some very significant developments. Three indicators – funding, publications, and patents – demonstrate that nanotechnology has actually been engaged in a flurry of infrastructure development and research activity.

The start of large-scale public investment in nanotechnology can be traced back to 2000, and the start of the USA's National Nanotechnology Initiative (NNI). This is a funding program which directs federal funding to a number of government agencies, institutes, and foundations. By 2009, the NNI had allocated almost US$ 10 billion to nanotechnology activities.

This situation is mirrored across much of the world. Japan invested Y86.5 billion (€562 million) in 2008 alone, and the European Union has also allocated a significant share of the 7th research Framework Programme to nanotechnology. Even as some countries begin to reduce their funding activities, others are only just beginning to ramp up. In 2007, Russia created an investment organization (RUSNANO) to which it has committed 130 billion roubles (€3 billion).

Public funding has employed a large number of researchers, with the consequence that scientific publications on nanotechnology-linked topics have increased substantially. In

1996, the number of journal articles that had been published about a nanotechnological topic was just under 5000. By 2006, this had increased to 34 000 [4]. This is a very significant increase in the body of knowledge about nanotechnology, indicative both of the amount of research work being carried out, and the amount of that work which involved novel, reportable discoveries.

A similar upsurge in the number of nanotechnology patents can be seen. From an annual figure of around 300 patents submitted to the European Patent Office in the 1990s, annual nanotechnology patent applications increased to 1000 in 2004 [5]. Nanotechnology patent applications continued to increase at a similar rate until 2007–8, at which point they showed a slight decline.

This is a typical pattern in technology development. Patenting activity first increases as initial discoveries made during fundamental research work are protected. This is followed by a phase of lower patenting activity during which the applicability of the research results is tested and commercially viable solutions are selected for further development. During the consequent application development phase patenting activity increases again.

In addition to funding nanotechnology research activities, public investment has been employed to develop a substantial nanotechnology infrastructure. The USA now has a National Nanotechnology Infrastructure Network with 12 centers spread across the country. These are equipped with the costly cleanroom facilities and tools that are necessary for nanotechnology development, and are established as user facilities which the research and industrial community can use.

It is arguably this very large-scale public investment that drove nanotechnology hype between 2000 and 2005, whereas the hyping of the Internet was a result of private investment and the response of the public markets. If one believes that public interest drives public sector investment strategy, the declining interest in nanotechnology may be reflected in a reduction of public funding in the coming years. There are already indications that whilst absolute funding will increase, rates of public funding growth will slow considerably.

Whilst this evidence suggests that, in terms of public perception, nanotechnology is in the hype cycle's Trough of Disillusionment, important indicators imply that in actuality, nanotechnology is entering a period in which it has significant impact. The current positioning of nanotechnology, relative to the hype cycle and the S-curve, is shown in Figure 9.3. The current phase of nanotechnology development is named Nanotech 1.0; this term will be explained in more detail in the next section.

9.5   Towards industrial reinvention

Nanotech 1.0 is an umbrella term for a number of nanotechnology applications that will lead to significant new products entering existing markets. The term Nanotech 1.0 is also used to suggest a step change from the nanotechnologies that are currently commercially available. Examples of these include stain-resistant coatings for clothing

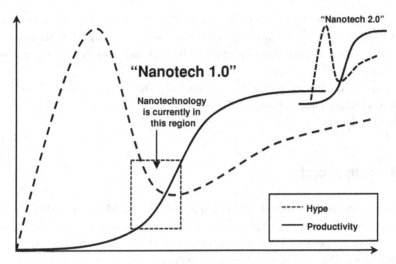

Figure 9.3 Positioning of nanotechnology, relative to the S-curve and the hype cycle (the latter adapted from [1]).

and hydrophobic wipes for car windscreens, which, whilst eye-catching and consumer-oriented, demonstrate little value creation.

The term Nanotech 1.0 is also used in order to distinguish this phase from future generations of nanotechnology that will have an even more widespread impact in a 5–15 year time horizon. Techniques such as self-assembly could transform manufacturing as we currently understand it and fundamentally alter existing industries. This phase of industrial reinvention would therefore be deserving of the title "Nanotech 2.0."

Examples of Nanotech 1.0 technologies are being seen in a range of industries, from energy to electronics and automotive. An example is atomic layer deposition (ALD), a technique which enables the creation of atomically precise thin films. Originally developed by Finland's Tuomo Suntola 30 years ago, ALD is currently used by Intel to produce high-k gate dielectrics for use in its 45 nm processors. The importance of ALD in this context is that it has been a key enabler in a technological breakthrough that Gordon Moore himself characterizes as the biggest change in transistor development in 40 years. These next-generation transistors are now foreseen to preserve Moore's law well into the next decade, while also enabling greater advances in energy efficiency.

Thin-film technologies are also being used to develop solar photovoltaic cells using cadmium telluride and copper indium gallium selenide (CdTe and CIGS). Major manufacturers including Nanosolar and First Solar are building plants which take advantage of the improved operational and production efficiency of thin-film solar cells. Thin-film solar technologies currently account for 10% of annual solar cell production.

The US-based company A123 Systems has built a business with over 2200 employees by developing a rechargeable battery which uses a nanostructured electrode. This technology was spun out from MIT and enables the creation of batteries which are capable of delivering higher power and shorter recharge times, and are far less volatile than current solutions. The company's batteries are currently being retrofitted to hybrid vehicles, transforming them into plug-in hybrids and reducing the amount of time for which the combustion engine needs to be used.

Each of these three technologies is enabling significant value creation for the companies involved and is having a wider impact on important societal challenges such as climate change.

9.6   Future hype cycles

Hype is inextricably linked with funding. A combination of high value creation investment and inflated values for publicly listed companies occurred simultaneously with the Internet's peak of inflated expectations in 1999–2001. The question should therefore be asked – will nanotechnology experience another hype cycle? One is unlikely to be caused by public funding, which will not experience the same rates of growth as previously seen. Instead, we may see a hype cycle linked to private investment in nanotechnology, whether through venture capital, industrial investment, or investment in listed nanotechnology companies. However, because nanotechnology is an enabling technology with multiple applications, this may actually be seen in a specific field or industry rather than nanotechnology *per se*. We may even have already seen such a phenomenon. Cleantech became a much-hyped investment class in 2008, and it could be suggested that this was a reflection of the technology developments – in solar, electric vehicles, fuel cells – that are enabled by nanotechnology.

9.7   Conclusions

The reality and perception of technology development often diverge. The example of the Internet reveals that whilst hype oscillates between fascination and despair, this often occurs whilst the significant, unglamorous work of technology development is continuing behind the scenes, as it were.

Nanotechnology is a further illustration of this point. Interest peaked midway through the 2000s, but whilst the wider community has become somewhat disenchanted, development has continued to a point at which nanotechnology is starting to have a significant economic impact. This is a consequence of the technology development that has occurred since the late 1990s – or since the late 1970s in the case of ALD – and that can be demonstrated with indicators such as patenting and publication activity. This corresponds to an upswing in the S-curve and is described as Nanotech 1.0, partly to distinguish these second-phase applications from the rather underwhelming nanotechnologies in early commercial products. These applications offer something genuinely valuable, beyond that provided by existing technologies, such as batteries with sufficient power density to make electric vehicles a practical alternative to the internal combustion engine.

We are now in the early stages of a period in which the promise of nanotechnology is beginning to be realized, and in which the large sums of public money that have been invested are beginning to bear fruit.

References

[1] J. Fenn, *When to Leap on the Hype Cycle*, Gartner Group, 1995.
[2] T. O'Reilly, What Is Web 2.0? Design Patterns and Business Models for the Next Generation of Software, 2005. Available at http://oreilly.com/web2/archive/what-is-web-20.html, accessed June 12, 2009.
[3] Google Trends, http://www.google.com/trends, accessed June 12, 2009.
[4] *The Nanotech Report*, fifth edition, Lux Research, 2008.
[5] C. Kallinger, *Nanotechnology at the European Patent Office*, European Patent Office, 2007.

10 Conclusions

T. Ryhänen

We have discussed mobile communication, the Internet, and nanotechnologies as a toolkit for building the next phase of human progress. All these technologies can be understood to have profound capabilities that will shape our economies and our everyday lives. Our specific focus has been on studying the impact of nanotechnologies on mobile devices and services. Nanotechnologies enable integration of new functionality into mobile devices and thus the creation of new digital services. We have also seen that nanotechnologies can be used to link digital information to various processes in the physical world, including social networks and urban infrastructures.

This book has discussed various key technologies: materials, energy, computing, sensing, actuation, displays, and wireless communication. We have identified potential technology disruptions for all these areas. Most nanotechnologies are still in the early stages of research, and the engineering effort to commercialize them will take at least 10–20 years. Their impact will be gradual: new nanomaterials will at first complement traditional solutions and improve the characteristics of various technologies. The disruptive nature of nanotechnologies will finally manifest itself by changes to our ways of interacting with the physical world, our environment, and our bodies.

The capability to tailor the properties of bulk and surface materials will affect our future handheld and wearable devices making them more robust, intelligent, transformable, sensitive, and environmentally friendly with functionalities embedded in these new materials themselves. Instead of using materials just to create functional devices, the materials will have meaningful integrated functions and even multiple simultaneous functions. These multifunctional materials with tuned, desired properties can be classified and defined as a library of functions to build future products. Bottom-up self-assembly and top-down fabrication processes need to be combined to create manufacturing solutions on an industrial scale. These nanomaterials and fabrication processes may also lead to more decentralized manufacturing solutions that can be brought closer to the consumer. Furthermore, we have the opportunity to develop manufacturing solutions that consume less energy and to create materials that are easier to recycle and handle throughout their lifecycle.

Nanotechnologies will improve the efficiency of energy sources and the capacity and performance of energy stores. The functionality of handheld and wearable electronics

Nanotechnologies for Future Mobile Devices, eds. T. Ryhänen, M. A. Uusitalo, O. Ikkala, and A. Kärkkäinen.
Published by Cambridge University Press. © Cambridge University Press 2010.

is limited by energy available and heat dissipation. Thus the energy solutions for future mobile devices need to be based on both improved energy storage and more energy-efficient computing. Furthermore, energy harvesting will be essential in creating various autonomous intelligent devices for ambient intelligence. We have shown concrete solutions for increasing the energy and power densities of batteries and for improving the efficiency of photovoltaic cells using nanostructured materials and devices. Many of these technologies are already being scaled up to support industrial production. The combination of nanotechnologies, mobile communication, and the Internet can also create global solutions to improve and optimize the energy efficiency of various processes of manufacturing and transportation, including the global information networks themselves.

The vision of future global information services is founded on the capability to handle, process, and store a huge amount of heterogeneous digital information at various levels of complexity throughout the "cloud," the network of shared computing capability. Ever since Richard Feynman's early visions, mass storage solutions have been among the most discussed application of nanoelectronics. The need for ever denser memory technologies is obvious, as mobile phones and portable multimedia players enable both the creation of the consumer's own digital content and the fast download of media, such as video, pictures, and music. We will soon carry terabytes of memory in our pockets based on nanoscale memory density.

Pervasive computing, sensing, and increased heterogenous digital information require more energy efficient computing solutions. Silicon integrated circuits based CMOS technologies will be able to fulfill the requirements of many emerging digital applications. Nanomaterials will have an essential role in the CMOS roadmap allowing improved material properties and smaller stable structures, e.g., the use of atomic layer deposition has already enabled the manufacture of very-high-quality insulating films. Eventually, the CMOS roadmap will reach its fundamental limits, which are set by increasing leakage currents and quantum mechanical tunneling. However, we have seen that electrical charge still forms a very competitive and energy-efficient state variable for Boolean electronics with the current types of architectures. Alternatives, such as spintronics, are still in an early research phase. The disruption in computing may be related to novel architectures for application-specific purposes, such as sensor signal processing and cognitive computing and to novel manufacturing processes for electronics.

The use of nanotechnologies in sensing and sensors has had an impact in three areas: new materials that improve the characteristics of sensors, detection of signals at the nanoscale with optimal nanoscale transducers, and functional arrays of nanoscale transducers and signal processing components. We have studied various mechanical, chemical, and optical sensors based on different physical or chemical nanoscale transduction principles. In general, miniaturization of transducers is not necessarily the best way to improve their performance, particularly in terms of resolution and stability. A good rule of thumb is that the size or impedance of the transducer needs to be matched to the size or impedance of the measured object. Thus nanoscale transducers improve the performance of chemical or biochemical sensors where measured objects are at the same molecular scale. We have also seen that we can learn from nature's ways of sensing, e.g., employing huge parallel arrays of sensory elements and signal processing using the same sensor array. Nanoscale components are also enabling the implementation of neural networks that consist of transducers and signal processing elements.

Nature has developed ways to create macroscopic movement and force based on incremental nanoscale movements, i.e., on protein nanomotors. Mammalian muscle tissue is the most refined example of such a system. There have been several attempts to mimic muscles in artificial systems. However, the design and optimization of actuators in terms of response time and maximum strain and stress depend on the requirements of the application. The design of artificial material structures, mechanical metamaterials, could be a way to optimize and tailor the right combination of properties for actuators. Mechanical metamaterials can also be used to embed into materials' mechanical functions, such as a preferred direction of bending, hinges, or locking mechanisms. Another interesting area is the use of nanoscale actuators, nanomotors, and nanorobots in manufacturing processes to enable the assembly of nanoelectronics and nanosystems. Actuation based on nanoscale mechanisms is currently limited to nanoscale movements but nature has already shown the potential for nanoscale motors driving actuation at the macroscopic scale.

Mobile communication and ambient intelligence are based on radios. Radio technologies have already developed miniaturized systems of silicon-based integrated circuits that can be embedded into most physical objects. The development of radio technologies and protocols for mobile communication has reached a stage where several competing systems (and sometimes several generations of the systems) exist in parallel. It has been challenging to balance the need to make mobile phones that are compatible with different radio and network protocols with the requirements for low power consumption and high overall device performance. During the coming 10–20 years radio systems will develop towards cognitive radio concepts where the radio has the capability to sense its radio environment, adapt, and dynamically allocate optimal radio bandwidths, and to learn from its user's behavior and the surrounding environment. Nanotechnologies are expected to enable energy-efficient spectral sensing and computing solutions. However, nanotechnologies will also enable embedding of radio communication into increasingly smaller systems.

Display technologies have been one of the key enablers in extending the communication capabilities of mobile devices. The quality and capabilities of the display are critical for this user experience. New consumer needs, i.e., pervasive access to the Internet, its content, and services, will drive the creation of new user interface concepts. Desired future user experience is an intuitive combination of multiple interaction modes: sound, visual images and video, text, natural language, gestures, and touch. Ultimately, the user experience will become a mixed or even immersive virtual reality. Displays of future devices need to have high resolution and contrast, consume less power, be flexible and conformable, and contain additional functionality, such as touch sensitivity and tactile feedback. Many commercially feasible display technologies are based on nanomaterials, such as liquid crystal polymers, e-inks, carbon nanotubes, organic molecules, and polymers in organic light emitting diodes (OLEDs). Furthermore, field emission using high-aspect-ratio carbon nanotubes and quantum dot light emitters are two examples where display concepts are directly based on nanoscale physical phenomena. Nanomaterials will also enable integration of additional functionality in displays, such as sensing, protective, self-cleaning, antireflection coatings, and energy harvesting.

Although the structure of this book is based on the key mobile device technologies, this has the clear disadvantage that this approach is not "future-proof." As we have seen in the discussion of the Morph device and technologies, nanotechnologies may lead to a completely new way to integrate different nanoscale functionalities. For example, transducers that are combined with signal processing and memory, and energy sources that are combined with radio or optical interfaces, will be integrated to give a huge parallel array of autonomous nanoscale systems. Thus the division of functionality will become blurred. In the face of this new type of integrated system, how will future electronic systems be partitioned?

The value networks of future electronics manufacturing will reflect these new ways of integrating electronics functionalities. Printed electronics, plastic electronics, and reel-to-reel manufacturing of tailored electronics are the first steps in this direction. Our global networked lifestyle will also affect the ways of organizing the manufacturing of future products. Nanotechnologies in combination with the Internet-based global value creation networks will be a source of extremely fast innovation that will correspond to the tailored needs of consumers. Our concept of innovation processes will change; open innovation will extend from software and the Internet to other areas of human productivity. We passed the nanotechnology hype phase some years ago; today we are seeing the first concrete nanotechnology products, but the real impact on industrial value networks is still some years ahead.

The stock market values of nanotechnology companies have dropped significantly. Also public interest in nanotechnologies has clearly declined. This indicates that nano-technologies have most probably passed their peak of inflated expectations in the sense of the Gartner hype cycle. On the other hand, there are examples of nanotechnologies that have already reached commercial maturity. Similarly, nanotechnologies have influenced several main stream technologies: integrated circuits, energy solutions, engineering materials, healthcare, etc. Following the analogy of the development of Internet business, we can call this phase Nanotech 1.0. Today we can already foresee a reinvention of the industry, Nanotech 2.0, in terms of larger-scale impact on various value networks and more tangible consumer benefit and value.

Finally, human creativity is required to find solutions to global population growth and environmental problems. Our greater capabilities to measure, gather, aggregate, and share global information, and our deeper capabilities to understand and master our interface to the physical world via bio- and nanotechnologies are the toolkit for building sustainable economies and lifestyles.

Index

Printed in the United States
by Baker & Taylor Publisher Services